The Family Smallholding

The Family Smallholding

KATIE THEAR

B. T. Batsford Ltd London

First published 1983

ISBN 0 7134 1935 0 (cased)
0 7134 1936 9 (limp)

Filmset in Monophoto Baskerville by
Latimer Trend & Company Ltd, Plymouth
and printed in Great Britain by
The Anchor Press Ltd
Tiptree, Essex
for the publishers
B. T. Batsford Ltd.
4 Fitzhardinge Street
London W1H 0AH

Frontispiece *The author outside Broad Leys*

CONTENTS

PREFACE

The aim of this book is to give practical information about the realities of living and working on a modern smallholding, illustrated by experiences of the author's family who moved to live on a two-acre site in the village of Widdington in Essex.

The book examines our motivations for moving to live in the country, the problems encountered and the order of priorities adopted. It is a personal record of how one smallholding is run, but the information is applicable to a wide range of differing situations and I have set our experiences against the broader spectrum of small farming generally.

The problem of earning a living is also considered, for it is unlikely that a smallholding will provide an adequate income, unless there is a high degree of specialization. If run efficiently, it will certainly reduce living costs by providing a large measure of home-produced food; it may also provide a useful source of supplementary income. In our case, the problem was solved by having a small business based on our smallholding. We spend half our time earning a living in this way, while the rest of the time is devoted to the land. As an example of this dual use of a smallholding, the birth and development of our business, the production of our magazine *Practical Self Sufficiency* is examined.

How to operate a smallholding so that it does not become a full-time operation, taking all one's time, is also an important question, particularly for those like ourselves who work at home to earn a living and who therefore do not have the time to devote all their attention to the land and livestock. We followed the same pattern as each new project was started. This was first to research the field, then, having acquired all the available information, to start the project ourselves. As problems emerged, we tried to learn from them so that we could adapt our techniques and evolve the best, most time- and labour-saving method for each enterprise. At the same time, we were concerned to avoid any degree of 'factory' farming and to keep our livestock humanely.

The Family Smallholding is therefore essentially a practical handbook, and is suitable for the needs of anyone considering buying a smallholding, or making greater use of an existing one.

While many of the administrative measures and authorities mentioned are British, equivalent bodies and regulations do exist in the USA, and there is a list of useful addresses for American readers on page 139. Subject to climatic variations, the basic principles underlying the book's philosophy apply just as much in the USA as they do in Britain.

Unless otherwise credited in the captions, all the photographs are from the author's collection. The drawings are by Jeremy Gower

All the vegetables at Broad Leys have been organically grown, without chemicals, for seven years

INTRODUCTION

In the Preface, the phrase 'modern smallholding' is used intentionally, for there is a big difference between the traditional smallholder, whose standard of living was often little above subsistence level, and a modern family living on a small farm. What is a modern smallholding?

Until comparatively recently, 'smallholding' had a specific definition. It was a small acreage of land rented out for agricultural or horticultural purposes. The landowner was usually a large farmer or one of the landed gentry, although local authorities and private or public organizations were also frequently involved. Smallholdings were usually run as family concerns, passing from one generation to the next, and forming an integral part of traditional rural life. The mixed farming and horticultural enterprises normally provided the sole source of family income.

During the last 40 years, the trend towards large-scale, mechanized and specialized farming has made many smallholdings go out of existence, because they were no longer viable. In many cases, land was put to more profitable use, not necessarily for farming. Local authorities, faced with increasing costs, sold off many of their smallholdings; in some cases, they are still doing so. Housing and industrial developments have subsequently appeared on some sites that were previously small farms.

Not all smallholdings have disappeared, of course. There are still some administered by local authorities, private landowners and organizations such as The Land Settlement Association. This is a body which was set up during the Depression years to administer government land, made available as smallholdings, for unemployed workers who wanted to start a new career in agriculture. Many such smallholdings now concentrate on glasshouse and other horticultural crops, and receive help in co-operative marketing and distribution. They are run by full-time, commercial smallholders.

In the last ten years, there has appeared a new phenomenon. Many thousands of people, often from the towns, have moved to live in the countryside. This is the first, small reversal in rural decline that has occurred since the Industrial Revolution. Although the numbers are small by comparison with the overall population, it is still a significant minority. What these families usually have in common is an interest in producing more of their own food, and in small-scale mixed farming conducted on a non-intensive and humane basis. They come from widely differing backgrounds, often have a full-time profession or part-time occupation in a nearby town and frequently run their own small businesses from home.

These 'new' smallholders differ from the traditional ones in that they normally own their land and have as their main income source an activity which is not associated with the land. They are sometimes regarded with suspicion by the farming fraternity, but, whatever individual attitudes may be, the reality of the situation is that there are far more part-time or 'private' smallholders than there are registered, commercial ones.

Many factors have contributed to this resurgence of small, part-time farming. Inner-city decay, urban violence and over-population generally have all played a part, as well as disillusionment with the materialistic nature of society. There is, perhaps, a feeling that the quality of life is higher in the country, where there is an emphasis on village institutions and neighbourly behaviour. Villages are not without their problems, of course, and the loss of services such as an adequate public transport system, local schools and village shops makes life in the countryside less convenient than in the cities or suburbs. At the same time, the new smallholders are having a rejuvenating effect on rural life. They are supplying villagers, who have lost their village shops, with free-range eggs, goats' milk and yoghurt, and a range of fresh, home-grown fruit and vegetables that are free of chemicals. They have also been instrumental in reviving the traditional practice of bartering. In Britain today, there are local self-sufficiency groups all over the country, where people meet regularly to barter or sell their surplus produce.

Our own move to the country, and our involvement with a smallholding, came as a result of various factors. I was born on a Welsh smallholding which had been in my family for generations. My mother ran the smallholding in the traditional way, with two cows, a few sheep and pigs, poultry and a large kitchen garden. It provided most of our food, and butter was made twice a week from the milk of the Welsh Black cows. We did not live solely from the smallholding, for my father, like many of the men in that part of North Wales, was a sailor, and had gone on his first voyage in a sailing ship called the Cambrian Princess, around Cape Horn, at the age of 14. It was a hard life, soon to be disrupted by a move to live in Liverpool, in search of better job opportunities.

After my schooling and a period of higher education, I spent several years teaching Biology and Rural Studies in Liverpool and London. Then I met and married my husband David in 1965. There followed a period of increasing material plenty. David had rapidly worked his way up to being an advertisement director for a firm of magazine publishers. Matthew was born in 1966 and Helen followed two years later. We led extremely comfortable and ordered lives.

In 1971 the blow fell. David was suddenly made redundant and we woke to find ourselves with large financial commitments and no money. It was in the days before large redundancy payments became the norm and all we had was a few hundred pounds. The company car went back and we were left with bicycles as the only means of personal transport. So, there we were, realizing for the first time how vulnerable we were to the tides of fortune.

There followed a difficult period in our lives. Many of those whom we had regarded as friends disappeared overnight. Redundancy was, at that time, still a rare enough event to result in social ostracism in some quarters. At the same time, we discovered the depths of real friendship in others. We found ourselves having to undergo a complete reversal of roles when I went back to full-time teaching, in order to bring in an income, while David stayed at home to look after the children. During this period, we were desperately trying to make some long-term plans. One quite deep conviction was that, whatever we did, we would start our own business and never again rely on someone else for a living.

The decision to start our own enterprise together was clear, but less clear was what it should be and how it should be financed. Despite various attempts at making extra money, we were still in financial straits and had to cut back on our spending. This meant fewer clothes, more purchases at the Oxfam shop, cheaper food and generally making do with less. We dug over half our lawn and grew fresh vegetables to supplement our diet. Eventually we were able to obtain an allotment nearby. We also kept rabbits for meat, a practice which had been common in the towns as well as the country during the war, but which had fallen out of use during the more prosperous years which followed. These activities were regarded with great suspicion by some of our neighbours, for the houses in our road all had beautifully tended lawns and rose beds. What few of them realized was that real need drove us to do these things.

While I continued teaching, David converted the spare bedroom into an office. We had decided that we would start a small publishing agency, selling advertising on a contract basis, for magazines which did not have their own, internal advertising departments. I would continue teaching until such time as the business showed signs of beginning to make a profit. David set off to Germany in a tiny, hired car and called on various magazine publishers. He had to give the impression of being a smart executive from a large agency, while in fact, he had barely enough money to eat. The venture paid off and he returned a week later with three contracts.

Eventually, I gave up teaching and joined David in the day-to-day running of the business. By 1973, we had not only come through the difficult period, but were enjoying a similar standard of living to that which we had before David's redundancy. Our attitudes, however, were quite different. We no longer took things for granted and found ourselves questioning why it was that so much of society was geared to materialism, at the expense of more important aspects of life. We had also, during the time when we were trying to produce more of our own food, realized how little practical information there was available to help people. Most of the gardening publications at that time were primarily concerned with decorative gardening, while the farming magazines catered only for the large-scale commercial farmer. There was nothing which told people how to look after half a dozen hens, an allotment, or rabbits for the table. *The Smallholder* magazine had ceased publication in the 1950s and nothing had replaced it. We began to wonder whether we could do something about this situation.

While David and our secretary bore the brunt of the day-to-day work of the business, I began to research the whole field of small-scale farming. This involved going on courses, poring over reference books in libraries and visiting small farms all over the country. During this period, I was also working as the UK representative for a European fashion magazine. This contract involved going to fashion shows and reporting on the trends for the journal. On one occasion, I arrived late at a fashion show, rushed in through the wrong door, and cannoned straight into a willowy model who was just gliding into one of those absurd poses. She ended up on the floor while I fell off the side of the demonstration platform, amid roars of laughter from the 30 or so journalists and representatives. I was obviously not cut out for great things in the world of fashion. This was perhaps just as well, for I could not help noticing the contrast between the worlds of small farming and fashion. Whenever I left a show and drove out to the country to talk to a smallholder, I felt that I was leaving an artificial and rather silly world and entering a more essential and real one.

At that time, I was also selling advertising, on a contract basis, for the British magazine, *The Ecologist.* This was an extremely valuable experience for it brought me into contact with eminent environmentalists such as the late Dr Schumacher, the author of *Small is Beautiful: A Study of Economics as if People Mattered.* The destructive effects of intensive farming methods on the countryside really came home to me, perhaps for the first time. Thousands of miles of hedges have been torn out since the war to make large fields for the intensive monoculture of arable crops such as barley. Natural wetlands have been drained, ancient forests felled and chemical spraying has had a devastating effect on indigenous wildlife. Many people in the cities still do not appreciate the massive extent of the damage that has been done, nor do they realize, for example, that it is far easier to keep bees in the town than it is in the arable areas of the country, because of the high incidence of sprays.

After a research period of approximately six months, we made the decision that we would publish our own magazine for smallholders and that it would concentrate on more natural farming and growing methods, and on keeping livestock humanely. At the same time, we decided to move to the country and run our own experimental smallholding. Our existing business did not need to be based in the suburbs and would transfer quite easily. There was no question of winding it up because we needed the income from it. It was also needed to support the magazine which we did not foresee as being a viable concern. Our aim was to produce a publication that would encourage people to write in about their experiences in order to help each other, and that would break even financially. We set out to look for a property and, about this time, I realized that I was pregnant. We were to have a new home and give birth to a baby and magazine in the same year.

We found Broad Leys in the small village of Widdington in north Essex and decided that it suited our needs. There was a house, a separate cottage called the Bothy which was destined to become the office, a few outbuildings and two acres of land. It was six miles away from the lovely old market town of Saffron Walden, with its fine timbered walls and traditional decorative plasterwork. The local schools had a good reputation, there was a quick train service to London which was 40 miles away and Cambridge, with its excellent cultural facilities, was only 16 miles in the other direction.

Reasonable travelling distance to towns and a large city is important, particularly for those who, like us, feel that concerts, plays and cultural development are an important part of life. The old fable of the town and country mouse is an interesting one, for it portrays the characters as being out of their depth and unable to cope when taken out of their usual environments. The town mouse is frightened by the country, while the country mouse cannot wait to get back home from the town. It has always seemed to me that the intelligent mouse is the one who experiences and therefore benefits from both worlds.

On 1 August, 1975, we moved in to our new home, but, by the time we had said goodbye to all our friends, it was the early hours of the morning. All our worldly goods were in a removals van which was not due to arrive until later that day. We crept out of the car – two adults (one heavily pregnant), two children, a dog, a cat and two rabbits – and promptly fell asleep on the floor of the cottage. We heard some distant quacks, but it was not until the following day that we realized that we had inherited a flock of Khaki Campbell ducks on the pond.

The business of earning a living had to continue and the Bothy was quickly converted into an office. We advertised for a secretary and an extremely capable 18-year-old came to work for us. Plans for the first issue of our new magazine continued and we made contact with a local printing firm to work out details of production. We had a budget of £400 to start the magazine which meant that we had to sell enough subscriptions to pay for the cost of the first issue. After that, all would be in the lap of the gods, but, in order to provide a financial back-up for the journal, we decided to start a mail-order book company which would operate through the magazine. David would run that while I would be responsible for the magazine.

We placed loose inserts in *The Ecologist* magazine and in the newsletter of The Henry Doubleday Research Association, a society which represents Britain's organic gardeners, and waited to see what the response would be. A single letter came in one day from a Mrs Prescott of Sydenham, together with a cheque for £3. She was our

Katie and David Thear with some of their raised beds
(Stephen Austin Newspapers Ltd)

first subscriber and with great delight we stuck her letter on the wall of the office. The next few days brought such a deluge of subscriptions that we had to abandon everything else and concentrate on writing names and addresses on cards.

Once the subscriptions began to come in, in early September, I felt panic for the first time. These people were seriously interested in the magazine, but it did not yet exist. The only thing I had ever edited before was the Parent-Teachers' Association newsletters of the children's school in Cheam. What was to go in the magazine? Lawrence Hills of The Henry Doubleday Research Association said he would write me an article on blackcurrants. John Seymour, the author of the best-selling book *Self-Sufficiency* sent in a piece called 'Back to the Land', and a craftsman friend, Chris Yates, produced a highly professional and illustrated article on sharpening tools. Dr Anthony Deavin, who was at that time doing research into aspects of organic growing, contributed what was to be one of the first articles on 'sprouting seeds' to appear in a British publication. He was also active on our behalf in getting other people to send in articles, including a *tour de force* on the Dexter cow. Several excellent and unsolicited articles appeared, as well as many letters of encouragement from subscribers.

Colin Richardson, a north-country man, had been producing a regular newsletter called *Self Reliance Newsletter*, which had a circulation of 600 subscribers. He asked us if we would like to take over the newsletter and incorporate it into our new magazine which we had, by now, decided to call *Practical Self Sufficiency*. We were delighted to do so, and the contributions he had received for the newsletter formed the beginning of a regular feature called 'Getting-It-Together'. This concentrated on readers' information, tips and suggestions, based on their experience. It proved to be one of the most popular items and still plays a major part in *Practical Self Sufficiency* today.

Soon the first issue of the magazine was complete, but there was to be no respite. I was already working on the second issue before the first issue was printed. In the meantime, I had made contact with a local doctor and midwife, and made arrangements for my baby to be born at Broad Leys. Gwilym was born on 12 November, 1975 – the first birth to have taken place at Broad Leys for over 100 years. He was a fine, healthy boy. While I was in labour I had been correcting page proofs for the second issue of the magazine. As each contraction came I put down my blue pencil and temporarily gave up reading.

The second baby, the first issue of *Practical Self Sufficiency* was born two weeks later on 26 November. A flood of letters poured in, full of encouragement and positive suggestions, as well as articles on every aspect of smallholding practice. There were a few of the other kind of course, including a tirade from a gentleman with problems who apparently did not approve of 'women farmers'. A retired military officer from Sussex accused me of leftist leanings and of encouraging hippies and drop-outs, while a left-wing intellectual wrote that I was catering only for the 'pampered middle-classes'. One of the nicest letters was from an old lady who had first kept chickens before World War One; her knowledge and experience were extensive, and typical of the high standard of material that came in from ordinary country people all over Britain.

Reaction from the media was far less positive. They were certainly interested in the new magazine, but only from the angle of 'a human interest' story. Attitudes have changed considerably since that time and it is interesting to see the more serious coverage which is now being given to small-scale farming. There is, at last, a realization that the phrase 'self-sufficiency' is only a convenient label to denote aspects of small livestock keeping and home growing; it has nothing to do with the classification of people. It soon became obvious to us that the people who were buying our magazine came from all walks of life, and were to be found in farming, in politics, law, education, industry, trade, the church and even the House of Lords. What such 'misfits and drop-outs from society' have in common, is an interest in livestock and in growing vegetables.

There is, of course, a lunatic fringe, just as there is in all walks of life, and the media must take the blame for portraying many of these, instead of concentrating on more ordinary people. 'Self-sufficiency' was portrayed as being living and dressing like Apache Indians in the Welsh hills, or wearing 'ethnic clothes' and eating brown rice while collecting supplementary benefit.

When, after the magazine was established, we began to hold Open Days at Broad Leys, we were able to meet many of our readers and to glean much valuable 'feedback' about the kind of material and information they wished to see in the magazine. They also provided an opportunity for people to get together and form local co-operative groups for such activities as discount buying of feedstuffs and equipment. The local community, in particular the crafts people and those with livestock, provided an enormous amount of help in the staging and organizing of these Open Days, as well as putting on practical displays of country crafts and farming and dairying skills. One of the remarkable things that we have noticed after each event, is that there is never any litter to pick up – despite the fact that several thousand people have spent the day in our back garden. This, we felt, was yet another indication that those with interest in this way of life are down-to-earth and practical people who respect the countryside and other people's property.

Such was the background to our interest in smallholdings and to our move to live and work on one. What I have not yet referred to is the system of priorities which we worked out so that the land, the livestock and the business could fit into an efficient and relatively harmonious whole.

1 The Priorities

When moving to a new house there are certain priorities which take precedence over other activities. When the home is also a smallholding, it is vital to ensure that the priorities are the correct ones.

The first essential is to concentrate on an adequate income. Where this comes from will vary, depending upon individuals, but without it, little is possible. The second priority is to ensure that the house itself is adequate. Repairs, renovations and extensions to provide comfortable accommodation take precedence over an extensive gardening programme or the acquisition of a motley assortment of animals. Outbuildings and fences are third in order of priority, and only when these have been repaired and appropriately adapted, is it prudent to think about keeping livestock.

As the priorities are so important, and because experience shows that there is a tendency to rush into trying to do everything at once, it is worth looking at these three aspects in more detail.

An Income in the Country

Some people have made the mistake of thinking that they can give up their jobs in the town, buy a smallholding in the country, and that this will, in some magical way, produce most of their essential needs. It is possible that if all the material blessings of the twentieth century were disposed of, someone adopting the subsistence life of a mediaeval peasant could survive, at least until bad weather conditions wiped out his harvest. I cannot conceive why anyone in his right mind would wish to try, unless it were for temporary experimental purposes or to win a large bet.

An income is essential for most people, but it need not necessarily be from a full-time job. Many people now living in rural areas are finding that a part-time job, in conjunction with their smallholding activities, produces a comfortable and satisfying lifestyle. The job may be working for someone else or operating a small home business. A business can operate from a smallholding, without necessarily having anything to do with agriculture, but not, of course, if it is a registered, commercial smallholding of the kind referred to earlier.

Many people are also finding that a smallholding can produce an additional income through local sales of surplus produce such as free-range eggs, goats' milk products or organically grown vegetables. In our particular case, our income comes from publishing. The smallholding activities are primarily for family needs, although we sell or barter produce that is surplus to requirements. Our business operates from home and there were certain regulations that we had to follow before we could start.

The Bothy, or small cottage near the house, had been occupied for domestic purposes. As this was to become our office, we had to apply to the local authority for 'change of use' planning permission. There was no difficulty about this because it was to be an office and there would be no noise problem. In a situation like this, neighbours are given the opportunity of lodging a complaint within a certain period after an application is made. Difficulties usually arise where the nature of a business is such that it causes noise or objectionable smells, which might be the case with heavy industrial use.

If a completely new building is required for a business, the local authority must again be approached, this time for actual planning permission, rather than just change of use permission. This is sometimes more difficult to achieve, particularly if an area is designated as a conservation area. Only individual local authorities are in a position to give specific advice on this. If permission is given for the erection of a new building, it is still necessary to acquire building permission. The latter is to ensure that minimum building safety regulations are adhered to.

Once local authority permission is obtained, a commercial rate will be levied on the premises. We pay domestic rates on Broad Leys itself, and commercial rates on the Bothy. It is worth remembering that, although costs of lighting, heating, equipping and maintaining business premises can be offset against tax, it may be necessary to pay capital gains tax in the event of the site being sold. An accountant will advise on this question as well as on any other problem that may arise with a home business. In fact, no business enterprise should be started without the advice of an accountant.

Our small Bothy office consisted of two rooms, a kitchen and bathroom with toilet. The building was in a poor state and needed fundamental repairs. The discarded wood provided useful fuel, but the old floorboards which were relatively undamaged were used in the

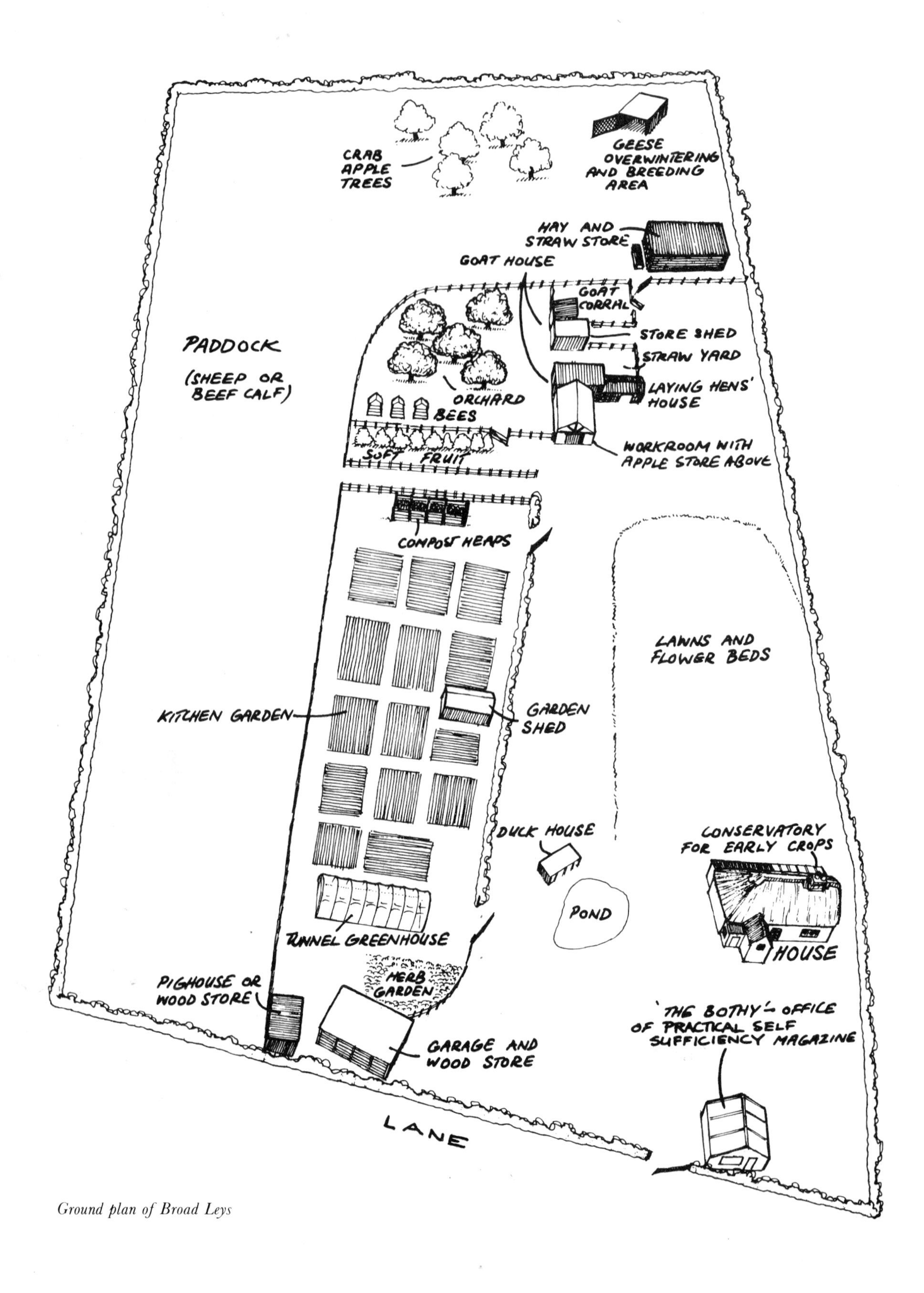

Ground plan of Broad Leys

subsequent construction of compost containers. Details of these are given in the kitchen garden chapter (page 24).

The two rooms of the cottage provided two interlinked offices, while the walls were utilized for storage. Once the desks, filing cabinets and typewriters were installed we could continue our business as before. The only snag we encountered was with the installation of a telephone for the office. As with many services in rural areas the waiting period was much longer than would normally have been the case in an urban area.

Seeing to the House

As mentioned earlier, there is sometimes a great temptation to acquire stock and start a gardening programme before adequate and comfortable living accommodation is established. In our case, there was an added incentive for ensuring that the house was a priority, for a new baby was due to arrive three months after we moved in. We saw the immediate needs as being electrical repairs and renovations, the provision of extra bedroom space, thatched roof repairs, kitchen modernization and general decoration. Each house will, of course, differ in its priorities, depending upon its condition and the needs of the occupants, but our particular experiences at Broad Leys may be of interest to those thinking of embarking on a similar venture.

The house is a timber-framed building, infilled with lath and plaster and topped with straw thatch. No one is sure how old it is. It appears to have grown in stages over the years from what was originally quite a compact cottage. County and parish records indicate that the site itself is extremely old and that a dwelling named Broad Leys has existed for many hundreds of years. The present house seems to have been largely rebuilt about 150 years ago, but using existing and much older materials. Many of the timbers are exposed and are clearly very old. One of the bedrooms has a mediaeval joint in the timber work.

On arrival, we inherited two downstairs rooms, a kitchen, storeroom, downstairs bathroom and toilet. Upstairs there were two double bedrooms and one single room. The roof leaked, the floorboards creaked and the wiring gave every impression of having been installed by Faraday.

Rewiring was an expensive job but it was obviously an urgent priority. A local electrician installed a double 13-amp ring main with plenty of points and including a safety trip switch. The flex, bakelite switches and weird assortment of plugs became things of the past and light came to Broad Leys.

'Just wait until the winter comes and you won't have lights,' said a neighbour cryptically. When winter came, the truth of this comment was revealed. The first high winds brought a power cut. We, like everyone else in the village, found ourselves lighting candles and waiting philosophically in the gloom. We had experienced power cuts in the town before, but there was usually a reason for them. Industrial disputes in the mid-1970s had produced power cuts all over the country. Now, we were to realize that, even at the best of times, the supply of electricity is a much more nebulous thing in villages than it is in urban areas.

The heating in the house was provided by an open fireplace in the living room, an oil burner in the hall and an ancient Aga in the kitchen. The open fireplace was really just a hole with a fire basket placed under a wide chimney. It consumed wood and coal with a voracious appetite and encouraged spine-chilling draughts from the ill-fitting Victorian windows. The oil heater was a Colman, possibly as old as the two-pin plugs had been. It was like a great saucepan in a metal box, with an oil burner at the bottom. To light it, you had to turn on the tap which brought in the oil supply from the leaky, old tank in the garden, and drop a lighted match inside. It was badly corroded and extremely inefficient. It gurgled and glugged with a considerable appetite for expensive oil and gave everyone headaches from the smell. During the winter, it was going day and night and we were still cold. It had to go. The priority, we decided, lay in getting rid of the Colman, installing a more efficient heating unit and insulating against draughts.

We made our own double glazing units for the north-facing windows. These were simple wooden frames fitted with 'trans-superglaze' polythene, a clear material, and held in place by thumb catches. They were effective in cutting out draughts and, in fact, worked well beyond our expectations. As they are light, easy to remove and store, it is a simple matter to remove them in late spring when the weather is warm, and to re-install them in the autumn. We installed similar ones in the office.

The roof needed no insulation, for thatch is like a giant teacosy, keeping warmth in in winter and providing cool conditions in summer. We have no loft and the bedroom ceilings soar up on a slant to the top. In some areas they are 4·2 m (14 ft) high.

We had to do something about the doors. Such is the peculiar nature of the house, that we have two front doors, one opening onto the much-frequented hall and the other going into the kitchen. Both were on the north-facing side and let in fierce, whistling draughts. The best long-term solution lay in providing porches for them both. As far as the kitchen entrance was concerned, this also made the storeroom an integral part of the house. Previously, there had been a ramshackle, lean-to affair where the dustbins were housed and which was always full of leaves blown in by the wind. It came down with a few hefty blows from a sledge-hammer, to be replaced by a porch measuring 2·4 m × 2·1 m (8 ft × 7 ft). The new porch not only provided us with an internal entrance to the storeroom, but gave excellent protection and insulation to the kitchen which, overnight, became draught-free and cosy. The second porch gave similar protection to the hall and to our official front door that no-one ever uses.

The Thear family during their first winter at Broad Leys, before the roof was rethatched and the extensions built

Below *Broad Leys with the roof rethatched and the verandah turned into a dining room and conservatory*

The Jøtul 118 space heater (Simon Thorpe Ltd)

We decided to replace the oil heater in the hall with a Scandinavian woodstove, a Jøtul 118. David had always been interested in woodstoves and had read widely about their design, uses and limitations, long before they were imported into Britain. Unlike other European countries, Britain, with its traditional reliance on coal, had little experience of using purpose-built and well-sealed appliances for the slow combustion of wood. We heard that a man named Simon Thorpe was thinking of importing Norwegian woodstoves into Britain and we immediately made contact with him in order to obtain one and to find out more about his operation. He became the agent for Jøtul in this country and set up a distribution system, selling many thousands of the stoves. Other people, realizing the potential, imported stoves from other manufacturers and countries and fairly soon British companies began to manufacture their own versions. In fact, there was something of a boom in woodstoves between 1976 and 1980, but it is Simon Thorpe who deserves the credit for introducing the concept to Britain and in doing the pioneering work

Below *Installation of a simple central heating system which does not rely on an electric pump*

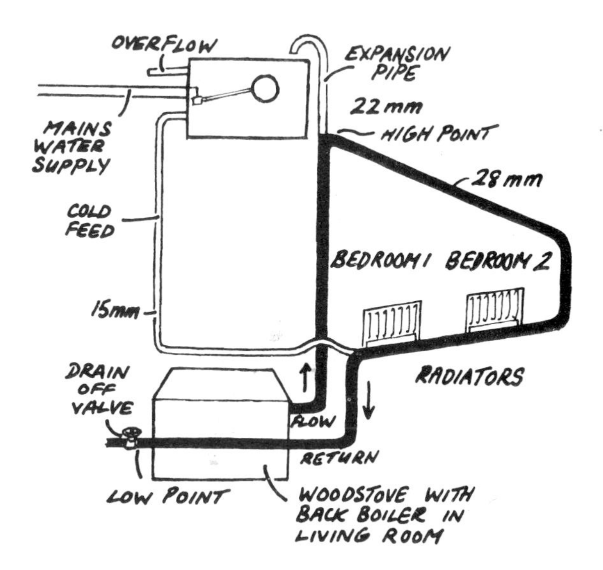

of educating people about woodstoves. He was tragically killed in a road accident.

Installation of the Jøtul 118 was straightforward, a matter of fixing the pipe into the chimney and blocking off the area around it. It stands on a stone flag, projecting out into the room so that heat emanates from it in all directions. It is essentially a large sealed, cast-iron box with draught controls and a tightly-fitting door. It takes 70 cm (27 in.) logs which burn slowly and completely from one end to the other, like giant cigars, and the heat transfer efficiency rate is high. Installed in the hall, it sends comforting glows of warmth in all directions, including up the stairs.

The open fireplace in the living room was replaced by another woodstove, a Kamina Mk 2 with a back boiler. This stove was chosen because it has a glass front so that you can see the fire, a factor we felt to be essential in the living room where we spend most of our evenings. The glass door can also be folded back into the hood for an open fire effect. The stove has a convector hood of steel over a cast-iron body. Air is drawn between the two, warmed and then rises. This convection system is excellent for warming up a cold room quickly.

The addition of a back-boiler to the Kamina was to feed radiators in the identical space above, which was the large double bedroom. This room was divided up into two separate rooms and a landing with store cupboard, by a local builder. The new rooms provided a single bedroom each for the two older children. Each had a radiator installed, connected to the woodstove back-boiler by 28 mm (1 in.) copper pipe. The design required one high point and needed to be a complete, simple loop. In this way, the heated water rose by convection and circulated through the radiators by gravity. It needed no electric pump and so was not at the mercy of winter power cuts. It is worth remembering that such a system is basically 'up and down', for, without a pump, the water cannot be persuaded to go very far sideways. In addition, the pipes need to be at least 15 mm ($\frac{1}{2}$ in.) and preferably, 28 mm (1 in.) in order to achieve the flow, and there must be a 'topping up' tank at a high point.

Wood needs to be stacked and stored under cover for at least a year, ideally two, before it is used, otherwise there will be tar deposited in the chimney. Under some circumstances this can 'creep' through brick or stone-work, creating an unsightly mess as well as a fire hazard. Wood is also a local and rural resource for it is bulky and expensive to transport. There is really no point in buying a woodstove unless a permanent source of reasonably priced wood is assured. Wood is the only renewable fuel on the planet, but, like all crops, it needs careful husbandry so that enough trees are planted on a regular basis for coppicing. The tragedy of Dutch Elm disease has provided fuel for a great many woodstoves in Britain but, in the long-term, alternative supplies will be needed. In our village we are particularly lucky, for the yearly coppicing of the surrounding wood, strictly controlled by local agencies, provides all the wood fuel for the village houses. It is simply a matter of placing the appropriate order in good time, specifying the length of log required and preparing covered storage space for them when they are delivered. One delightful result of having wood stacks is that they attract wrens and tree creepers who find all sorts of delectable insects amongst the logs. One year, we had a family of stoats nesting there too.

We had achieved light and warmth at Broad Leys but what about the thatch? A few weeks after we moved in in 1975, the suspicious-looking brown patch on the kitchen ceiling proved to be what we suspected: a place where water came in whenever it rained. Fortunately it was above the sink, but it was a nuisance. There was a long, sloping hip on one side of the house and through this thatched slope there soared a tall, brick chimney which carried away the exhaust fumes from the Aga. Immediately below the chimney, the water collected in a narrow channel and eventually worked its way through the thatch to plop into the kitchen sink.

We immediately contacted the thatcher, thinking that we could put up with the dripping water for a week or two, or possibly even for a month or so if he was particularly busy. How naïve we were about the mysteries of thatching supply and demand.

The thatcher told us on the telephone that he did not need to come and examine the thatch for he knew the precise condition of it, just as he knew every other thatched building in the district. He explained that he would come and see us in two years' time and give us an estimate then, because that would be the first occasion when he would be free. We had to reconcile ourselves to plopping water for some time to come.

We could, of course, have had the whole thatch stripped off and replaced with tiles, but we would have lost the insulation value and, just as important to us, the aesthetic appeal of the house. A large number of thatched roofs have been lost over the years. Once gone, they are rarely replaced and yet another part of our national heritage has gone. There is a widely-held belief that it is only the wealthy who can afford to live in thatched houses. This is quite untrue. Most of the thatched houses in our area, and there are many of them, are inhabited by ordinary families of modest means. It is true that house insurance premiums to cover fire risk are higher than normal, but there are also specialist thatch insurers who offer cover at reasonable rates.

True to his word, the thatcher appeared two years later, gave us an estimate and said he could do the work in the summer of 1978. It involved two new hips or side slopes, a replacement layer for the south side and a new ridge. The north side was virtually undamaged. Thatched roofs deteriorate much more quickly on the south-facing side because of the effects of the sun. We were relieved to find that the damage was not more extensive. At that time, like many people, we thought that having a roof rethatched meant literally taking off the whole roof and putting on a new one. In fact, this is rarely the case. Most roofs only require the top portions to be removed (the depth depending on the degree of damage), having new material inserted and then the whole thing 'combed' and netted. The wire netting is important because it ensures that birds do not get into the straw to nest. Our old roof had an extensive population of twittering fledglings, particularly in the section of roof above Matthew's window. In fact, once the roof was rethatched, he said that he missed the familiar sound.

Ever since we had moved in, we had wanted to re-do the kitchen, but nothing was possible until the thatching was done. The kitchen was old and inefficient and the Aga, which was a warm friend in winter, was also cantankerous and unreliable. It is not pleasant to cook a meal and then, on opening the oven door, to be met by a rush of yellow, sulphurous fumes. The cooker had a bad crack in it and the thermostat did not work. The Aga had come to the end of its useful life, but what was to replace it?

The choice lay between another solid fuel cooker or an electric one. Like many small villages, Widdington does not have mains gas available. The question of the chimney which had originally caused the leak in the thatch was also a vital one. It had missing bricks, leant to one side and was clearly unsafe. We decided to remove it entirely and cook by electricity. Personal preference as well as convenience played an important part in this decision. I enjoy cooking but I also appreciate convenience and modern technology. In the event of a power cut, we could fall back on the woodstoves as a source of heat for cooking.

Down came the chimney, felled by local builders. Everyone stood clear, but it was not such a dramatic event as a tree coming down; just a quick slither of bricks down the thatch until they landed on the grass. A few fell in the pond and caused a miniature tidal wave which set the ducks rocking from side to side in quacking protest.

The very next day the golden wheat straw for the thatch arrived, to be dumped without ceremony on the lawn. Thatching is a wonderful craft to see in action. It has remained virtually unchanged over the centuries, with many of the same tools being used from one generation to the next. The father and grandfather of our thatcher, Mr Osborne, were also thatchers and the family skill was obvious to see. In two and a half weeks, having been blessed with perfect weather, the task was complete. The thatch was 1·2 m (4 ft) thick at the base of the central chimney, shading down to approximately 0·6 m (2 ft) at the eaves. Where the secondary Aga chimney had been, there was now a beautiful and unbroken golden sweep.

We redesigned the kitchen to meet our needs and began by ripping out everything so that we were left with an empty room. The space was not big, but sufficient provided it was utilized properly. We put in a completely modern kitchen with continuous working surfaces, storage areas, built-in cooker, hob unit, refrigerator and dishwasher. In the storeroom, on the other side of the kitchen porch, was a large deep freeze, automatic washing machine and tumble dryer. The old Aga had proved stubborn to the end and had to be finished off with a sledge-hammer. All that remains of it now is the inner cast-iron pot from the firebox which is still in service for the forcing of early rhubarb in the kitchen garden.

It was ironic that, just as we had finished decorating the kitchen, I received a letter from a reader of the magazine saying that he and his wife had succeeded in cutting themselves off from the 'rat-race' of modern life, and had disposed of all the 'twentieth-century gadgetry, including washing machine and electric kettle'. He gave a long list of all the things that they were now doing without and my heart went out in sympathy to his poor wife, for they were all things that would have made life easier for her. I could not help noticing that there were very few items which applied to him, but he presumably did not have to wash clothes by pounding them with stones at the river bank.

There is undoubtedly a lunatic fringe associated with aspects of self-sufficiency, which seeks to turn back the clock and live a kind of peasant existence. This attitude seems to emanate primarily from discontented urban dwellers who have a nostalgic and totally unrealistic picture of rural life. I have never understood why the benefits of modern technology, which have done so much to ease the domestic burden of women, should be regarded as being in some way unwholesome. My dishwasher, in particular, has been a boon.

The final improvement we made to the house was, perhaps, our most ambitious project. It was to build an extension along the whole of the south-facing side of the house. One half was to be a dining room while the other would be a conservatory for the production of early and out-of-season food crops.

The house appears never to have had a dining room and our meals were held in the 'Piccadilly Circus' hall. For a family of five, with frequent visitors, this was unsatisfactory. We needed to spread out a little. The verandah itself consisted of brick pillars, a wooden frame and a corrugated, translucent plastic roof which leaked. The floor consisted of pink stone slabs. The side of the verandah which had the kitchen window opening onto it was destined to be the dining room, while the other half was to be the conservatory with access from the hall. Planning permission was obtained from the local authority, after only a brief period of waiting, and the work began.

An oil-fired central heating radiator was put in the dining room, kitchen, conservatory and east bedroom with a boiler housed in a small room off the porch by our unused front door. This also provides our hot water in winter. In the summer months, the latter comes from an immersion heater. The extra insulation provided by the dining room and conservatory means that overall heat loss is minimal and our heating bills, which are for a combination of oil and wood heating, are comparatively low. In winter the conservatory roof, windows and door are double-glazed with polythene. This is simply a matter of attaching it to the framework with small pieces of cardboard and staples. The squares of card ensure that the plastic does not tear. In spring, these are removed. Further details of the conservatory are given in the section on protected crops (page 42).

The hall, which was now completely enclosed and had no external walls, had to be made lighter. This was achieved by putting in a large window between the hall and dining room and replacing the solid doors with half-glazed ones. The overall effect is that the hall is now no darker than it was before we made the changes.

With the construction of the dining room and conservatory, all major renovations and improvements were complete. The only other work which we undertook was to make a patio on the south side of the house. As the garden slopes upwards, away from the house, this necessitated digging out a great deal of earth. A local farmer, armed with a mechanical shovel, did this for us, and dumped the earth on the north side of the house, some distance from the front gate. The pile was turned into a rockery which, at the same time, provided a much-needed windbreak and weather protection for that area of the garden. Previous experience of snow blizzards had made us realize that there was nothing to stop huge drifts piling up there, and we live in an area where snow is frequent and can be heavy in winter.

It is amusing to remember that, when the earth was first dumped, the rumour swept around the village that we were constructing a nuclear fall-out shelter. I was even telephoned by a national magazine, asking if I would write an article about its construction. When I explained that it was really a patio and a rockery that was involved, they inexplicably lost interest.

Outbuildings and Preparing for Stock

Outbuildings are essential on any smallholding. They are needed for housing livestock, storing hay, straw, feedstuffs, vehicles and equipment and, in some cases, may have a specialized function such as providing dairying facilities. Anyone considering buying a smallholding would be well advised to concentrate on one which has existing outbuildings in a good state of repair. The cost of putting up new buildings can be extremely high, and although grants may be available where a building can be shown to be for commercial farming or horticultural use, they are not readily available to the small, part-time farmer. The divisional office of the Ministry of Agriculture will give advice on this. The address and telephone number can be located in the local Yellow Pages directory.

Our outbuildings included a substantial wooden stable in the field. This had been used for horses, but we decided that it would be just right for storing hay and straw. It is important that these have dry, airy conditions such as those provided by the traditional Dutch barn with its roof and open sides. Hay, in particular, is dangerous if it becomes damp. It can, on rare occasions, ignite spontaneously because of the high temperature brought about by decomposition. If it is mouldy, it may give off spores of a fungus causing the disease Aspergillosis. This affects not only livestock, but also humans who may handle the hay. The alternative name for the disease is 'farmer's lung'.

At the bottom of the garden, some way from the house, there is a separate range of outbuildings at Broad Leys. When we arrived, these consisted of a double garage with an overhead storeroom; access to this upper room was via wooden steps on the outside. It is a substantial building made of brick and timber which has a tiled roof with an attractive weather vane on the top. We decided to remove the double garage doors and replace them with a single door and a wall with a large window so that the building could be used for storage and as a general workroom. A more convenient double garage was then constructed at the front and to one side of the house and adjacent to the lane. This was made big enough to allow for the stacking of logs at the back and down one side without interfering with cars. The upstairs storeroom, above the old garage, became a fruit store for keeping apples and pears through the autumn and winter.

To one side of the old garage, which is now a workroom, there were three small stables, all of which had been used to house horses. One of them had a concrete floor and a drain so we decided that this would make a suitable milking parlour. It is a great advantage to have a specific place where a dairy animal can be taken for milking, and where the floor can be regularly hosed down. If the building can be kept clean milk is less likely to be contaminated by dust or dirt, an important aspect of dairy hygiene. The second stable was divided into two goat pens and a feed store, while the third was eventually adapted as housing for winter-laying hens. Further details of these are given in the appropriate chapters.

The only other building on the site, apart from a garden shed and a few smaller storerooms, was the pump house. This was a brick building which had once housed machinery for a pump to draw water from a bore-hole and drive it into the house. When the property received a mains water supply, the old system was discontinued and the machinery removed. We turned it into a pig house which doubled as a subsidiary wood store when we were not keeping pigs.

Many people new to smallholding life do not realize that there is usually a turnover in stock and that a particular type of livestock may not be kept all the time. Because of this, buildings may have different uses at different times. This is particularly true where the acreage is limited, and animals must therefore be kept in rotation. As we have only two acres in all, we keep

The disused garage at Broad Leys, converted into a downstairs workroom and an upstairs apple store

Buildings for livestock need periodic maintenance, particularly replacement of the roofing material to ensure dry conditions (Ruberoid Building Products)

A tractor such as this second-hand Ferguson is useful for the larger acreage

milking goats permanently, but a beef calf, sheep and pigs are kept periodically. In other words, a beef calf might be reared on the grass for one year; another year, a couple of lambs would be kept instead. There would never be an occasion when all these types of livestock would be kept at the same time. This system not only protects the land from over-use, but ensures that we do not have more animals than we can cope with.

Before any stock is acquired, however, it is necessary to ensure that all fencing is adequate. A good, thick, prickly hedge will normally confine all livestock except goats, which will happily eat their way through the thorniest of barriers. In fact, it may be necessary to resort to tethering goats if the fencing is inadequate.

Electric fencing will confine most stock, including goats, and is particularly useful for a number of reasons. It is easily transportable and so can be used to confine stock in a particular area for a short time, before they are moved on to new grazing. This method of selective grazing ensures that one area is completely eaten down before a new section is made available, and is an excellent way of making the most economic use of grass. It is also useful for confining stock such as pigs in a part of the vegetable garden where winter digging is required. Pigs are natural ploughs and will turn over even the most difficult of soils, devouring roots as they go. The secret of success is to keep the confined area small so that the rooting activities are complete before a new piece is fenced off and made available.

Post and rail fences up to 90 cm (3 ft) in height will confine most cows, adult milking goats and horses. Bullocks and frisky goatlings may require a 1·5 m (5 ft) fence. Sheep and pigs tend to push through a barrier rather than attempt to go over and one of 90 cm (3 ft) height is normally sufficient, provided it is stable. Purpose-made sheep netting and pig netting is available and this needs to be well anchored into the ground. Some of the long-legged, more traditional breeds of sheep such as the Jacob may require a height of 1·5 m (5 ft) to confine them, because of their tendency to leap.

Stone walls make excellent barriers, but these are traditionally found in the north and western areas of Britain, where natural stone is available. Building and repairing such walls is a skilled occupation.

Our fences at Broad Leys are a combination of natural hedge, post and rail and netting. We also use an electric fence where selective grazing or some other controlled grazing is required. We were particularly anxious to ensure that any stock we acquired did not wander off our land and encroach on any neighbour's territory, especially as one neighbour had a beautiful three-acre rose garden, a kind of Elysian field as far as goats are concerned. The old adage of 'good fences make good neighbours' is very true.

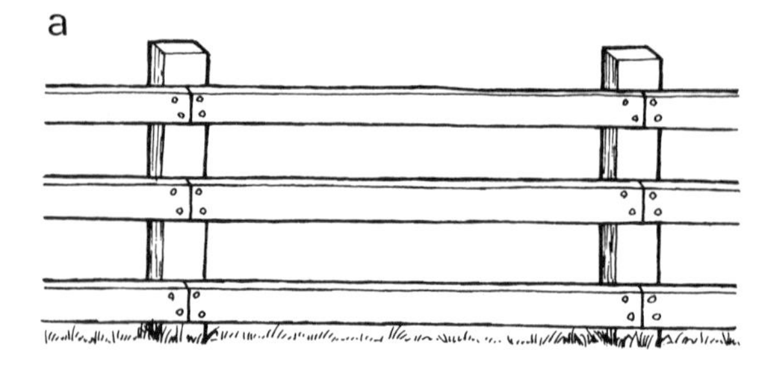

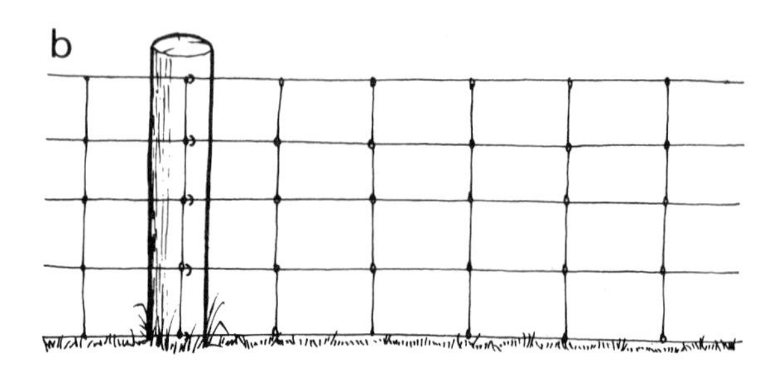

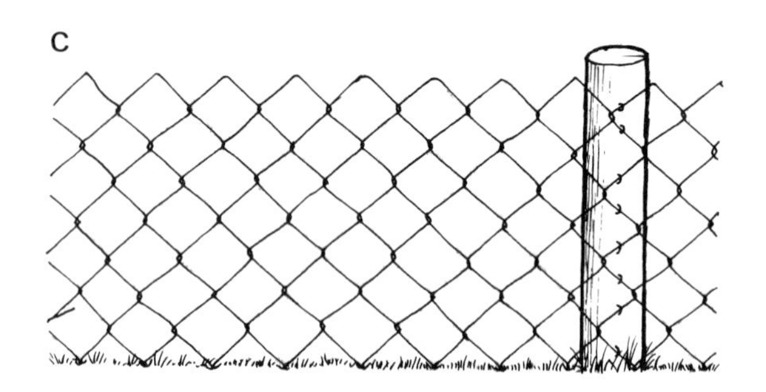

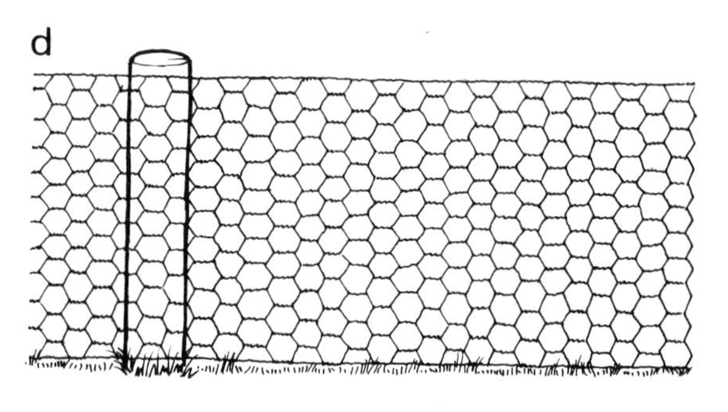

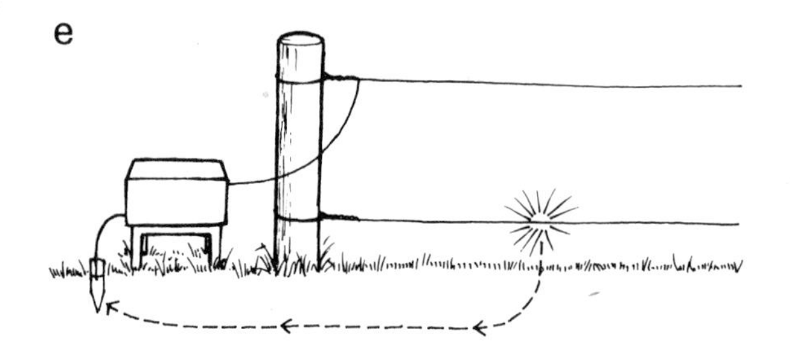

Types of fencing:
(a) Post and rail *(b) Pig netting* *(c) Sheep netting* *(d) Poultry netting* *(e) Electric fencing*

2 The Kitchen Garden

Our kitchen garden has to do several things. It has to provide most of our basic fruit and vegetable needs, with a surplus for preserving, and it is also expected to yield fresh food throughout the winter.

As a leisure activity, gardening is a healthy and interesting hobby but it can easily become a burden, imposing a strict regime of digging, weeding, planting, watering and more weeding, leaving little time for leisure. We wanted our gardening to be pleasurable as well as useful, so we looked for ways in which we could save time.

The Basic Plan

The priority lay in planning, which meant first deciding what to grow. This is not as obvious as it sounds. In a family there is usually a string of likes and dislikes. Children often have a tendency to regard being made to eat anything fresh from the garden as a violation of their human rights, preferring something atrocious from a tin. This sometimes proved embarrassing, particularly on the occasion when a lady from one of the women's magazines came to lunch and to do an interview. My *oeuf florentine* was a speciality of the house, with home-grown spinach, fresh eggs and home-made Cheddar cheese in the sauce. It was exactly what the media lady expected to see, but our youngest child, taking one look at my creation, wrinkled his nose in disgust and asked for a bag of crisps instead. So are the best-laid public relations plans brought to ruin.

Not to be deflected from our plan, we persevered in establishing what everyone would be likely to eat and simply did not grow those vegetables which were unpopular. As far as our family is concerned, this excludes swedes, scorzonera and turnips. Celery is not popular either but as I am fond of it and the new, self-blanching varieties are much easier to grow than the older types, we grow a small patch of it.

Our second decision was that the crops would be grown naturally, without resorting to chemicals. With livestock, there is plenty of manure available and once this is composted and rotted down, it provides an excellent boost to the soil fertility, avoiding the need to buy fertilizers. We wished to avoid chemical pesticides as well, for a number of reasons. They are expensive and once you start using them you are quickly trapped in an escalating spiral. Pests show great adaptability in developing resistance to chemicals, as the case of the glasshouse whitefly demonstrates. The whitefly has developed such a degree of immunity that there is now no chemical which can be used against it which will not also have a detrimental effect on the crops it is supposed to be protecting. Because of this, commercial growers of glasshouse crops are increasingly going over to safer, biological control methods. The most popular of these is the mini-wasp, *Encarsia formosa*, which is parasitic on whitefly and effectively controls the numbers.

Growing crops without using chemicals is not without its problems of course, but much can be done to avoid trouble. As a rule, early varieties are more free of virus and insect pests than later ones, although they are often at greater risk from hungry birds – but netting will provide protection in this case. It is generally the maincrop varieties of peas rather than the early ones which are affected by pea moth caterpillars. Similarly, early varieties of potatoes usually escape blight. This brought us to our third decision, which was, as far as possible, to grow those varieties which were disease-resistant. When a variety is so described, it does not mean that it is absolutely safe against a particular disease, merely that it has a greater resistance to it than other varieties. As an example of this, the Avonresister parsnip is more resistant to canker, an infection leading to nasty brown patches, than the older variety, Hollow Crown, so it seemed sensible to grow the former rather than the latter.

On the question of varieties, we also took flavour into consideration and concentrated on those which had a particularly good flavour. This meant that some commercial varieties were discarded on the grounds that, although the yield was high and the produce of a uniform size, the flavour was poor. Moneymaker tomatoes were discarded in favour of the Pixie variety because of this.

Choice of varieties is, of course, a personal one and it is essential to study the seed catalogues every winter to see what is available. It is always interesting to try out new varieties as an experiment, as long as you remember to keep records. In fact, keeping records is a good idea for all varieties, new and old. I use a hard-backed exercise book for mine and despite having muddy thumbprints and the occasional blank page torn out for my youngest to draw on, it is a valuable source of information for the next season. A sample page has the following headings:

RECORD							
Vegetable	*Variety*	*Date*		*Amount sown or planted (row length or number of plants)*	*Date(s) harvested*	*Amounts harvested (number or weight)*	*Comments*
		Sown	*Planted out*				

Our fourth decision was to concentrate on vegetables and varieties which were suitable for our particular soil, boulder clay overlaying chalk. This meant that the brassica family had almost ideal conditions, for this particular combination is to their liking, but maincrop carrots which like a loose, sandy or friable soil were at a disadvantage. At first, because of this, we concentrated on the small, early carrots and avoided the maincrop ones; but after we had experimented with raised beds, we subsequently grew these as well. Further details of the raised beds are given later in the chapter.

The fifth decision we made was related to yield of crop in relation to ground used. Some crops such as runner beans which grow upwards produce a heavy yield and yet occupy a comparatively small area of ground. This factor is crucial for those who have only a small growing area. It was also important to us because, although the overall site amounts to two acres (0·4 hectares), we did not want to extend the kitchen garden at the expense of pasture area for the livestock. In 1976 we carried out an experiment to discover what yields were possible from a given space so that we would have a better indication of what was worth growing on our soil. A trial plot 3 m × 6 m (10 ft × 20 ft) was dug over, manured and regularly tended and watered. Each row was the width of the plot – 3 m (10 ft). The yields were as follows:

2 rows early peas 2·3 kg (5 lb)
2 rows maincrop peas 2·7 kg (6 lb)
double row of runner beans 9·2 kg (20 lb)
2 rows of dwarf (French) beans 5·4 kg (12 lb)
1 row summer cabbage 12 heads
1 row summer cauliflower 8 heads
1 row Brussels sprouts (5 plants) 2·7 kg (6 lb)
1 row beetroot 4·6 kg (10 lb)
1 row early carrots 1·1 kg (2½ lb)
2 rows onions (sets) 9·2 kg (20 lb)
2 rows leeks 10·9 kg (24 lb)
lettuce (in odd places) 30
salad onions (in odd places) approx. 4 bunches
radish (in odd places) approx. 6 bunches

These yields indicated to us that some vegetables are more worth growing than others. For example, the runner beans, dwarf beans, cabbages, beetroot, onions, leeks and salad vegetables produce good yields from relatively confined spaces. Peas and sprouts, on the other hand, require a lot of space for comparatively low yields.

We took the decision not to grow some maincrop varieties, notably potatoes and peas. Mention has already been made of the tendency for these to be affected by pests. In our particular area, maincrop potatoes are available cheaply, by the sack, from local farmers and there is also a large pea-growing concern which markets freshly picked and podded peas all ready for freezing. Although these are not organically grown, we felt that it was a sensible compromise to do this, while growing the early potatoes and peas organically. We also decided that it was worth growing sprouts because, although the yield in relation to ground occupied was fairly low, it is one of the best winter vegetables when there is little else available and it is popular with all the family.

To summarize our basic plan for the kitchen garden, we had decided to grow the following:

- crops which the family would eat
- crops which could be grown organically without too much difficulty
- varieties which had good flavour and disease resistance
- crops which were suitable for the particular soil and conditions
- those that produced a good yield for the amount of ground occupied and which could not be bought cheaply elsewhere

So much for the planning, but the most daunting of tasks was getting the ground in a suitable state for cultivation.

Preparing the Ground

Some gardening books seem to assume that one starts with an area of bare ground and all that is necessary is to turn it over and leave the weather to break down the soil into that other hypothetical dream, the fine tilth. We, like many others before us, were faced with the grim reality that there is no such thing as an expanse of bare soil just waiting to be cultivated. 'Nature abhors a vacuum' and any piece of earth exposed to the air will quickly attract a myriad of weed seedlings, all competing to produce a tangled mass of undergrowth. Many people, when renting their first allotment, have discovered to their dismay that they have inherited a solid mass of thistles, nettles, docks and couch grass, with a waving mass of growth above ground and a dense network of deep roots below. To try and dig this type of ground is obviously out of the question. The top growth must first be removed before the roots can be attacked. It is only then that the soil itself can be turned.

Our particular kitchen garden was created from a portion of paddock which meant that the top growth of grass had to be skimmed off before anything was possible. The area was marked off and then the turf cut, square by square, and skimmed off with a sharp spade. Where nettles, docks and other tall growing weeds were encountered, these were cut at ground level so as not to interfere with the initial skimming. The turf was stacked, grass side down so that it would eventually rot down into loam soil. We then turned our attention to the bare earth and made a disastrous mistake.

It had seemed obvious to hire a mechanical cultivator and turn the ground over mechanically. As it happens, this is a good idea, but we neglected first to dig out the roots of perennial weeds. What the rotovator did was to chop up the roots of the weeds into small pieces and distribute them all over the site. Each small piece of root grew into a new plant.

Digging perennial weeds by hand is hard work and time-consuming, but is well worth doing in the long term. The area can, of course, be treated chemically to kill off the roots, but many people are concerned at the possible effects of the toxic residues in the soil. These can have a detrimental effect on wildlife, on earthworms and the micro-organism population of the soil – and possibly on those who eat the vegetables grown on the treated soil.

Many organic gardeners have found that a compromise is possible by using a non-toxic weed killer such as ammonium sulphamate. This is available under the trade name of Amcide and is effective against couch grass, docks, marestail, comfrey and oxalis but not against nettles and thistles. It is non-selective so the ground must be left for six months before anything is planted there, but meanwhile, it gradually decays into sulphate of ammonia which acts as a fertilizer. 455 gm dissolved in $4\frac{1}{2}$ litres of water is enough to treat 9 sq. m of ground (1 lb in 1 gallon for 100 sq. ft).

A cultivator cuts down the work of ground preparation (Wolseley Webb Ltd)

Once perennial weeds have been removed, either by hand or chemically, the ground can be brought into cultivation. If a rotovator is used, and this is certainly the quickest way, well rotted manure or compost should be spread over the surface first. As the rotovator turns the soil, the manure is incorporated so that the fertility of the soil is increased. It is important to ensure that the manure or compost is completely rotted (see below) otherwise nitrogen depletion will occur – nitrogen can be actually stolen from the soil as the decomposition process of the manure continues.

Digging by hand will obviously be slower than mechanical rotovating, but there are many who claim that the resulting soil is better because cultivation by spade is generally deeper than is possible with rotovator blades. As digging proceeds, rotted manure or compost can be added. How this is done is largely a matter of personal preference. It can be spread over the whole surface first so that it is incorporated as you go, or it can be forked from a wheelbarrow into the particular trench on which you are working.

At some point, lime will need to be added to the soil, particularly if it is fairly heavy clay, but it is important not to apply it at the same time as manure otherwise there will be a reaction between the two and the fertility will be decreased. If the initial digging or rotovating is done in the winter, the lime can be added at this time, while the manure is left until the spring. Where the soil is a clay one, the lime has a flocculating effect in that it helps to break down solid lumps of clay, redistributes individual clay particles and improves the structure of the soil. The effect of frost on roughly dug heavy soil will also help to break down the clods. Lime is also needed to 'unlock' other elements in the soil which might otherwise remain unavailable to the plants. It counteracts acidity and has a general 'sweetening' effect on the soil, as well as making the earth more attractive to the earthworm population. The level of acidity or alkalinity of a soil is referred to as its pH value. A pH of 6·5 is neutral. This can be measured by taking soil samples and conducting a simple test on them. Soil-testing kits are readily available in garden suppliers' shops.

Increasing the Fertility

Unless bought-in inorganic fertilizers are being used, it will be necessary to maintain the fertility of the soil in other ways. Crops will use up the available nutrients and these must be replenished if reduced cropping and damage to the soil structure is to be avoided. Where livestock is kept, there will be a plentiful source of organic matter, but it needs to be stacked and allowed to rot down completely before being added to the soil. A convenient way of doing this is to build compost containers which will not only keep the stack in a tidy condition, but will also ensure adequate aeration and warmth to accelerate the breaking down process.

Our compost containers were built of recycled materials, including the old floorboards and wall panelling of our Bothy cottage, discarded when renovations were carried out. Wooden posts were sunk into the ground to make the four corners of a box. Three sides of this were made by nailing wooden boards across, leaving small gaps for adequate aeration. The front had battens nailed to the two uprights so that a gate structure

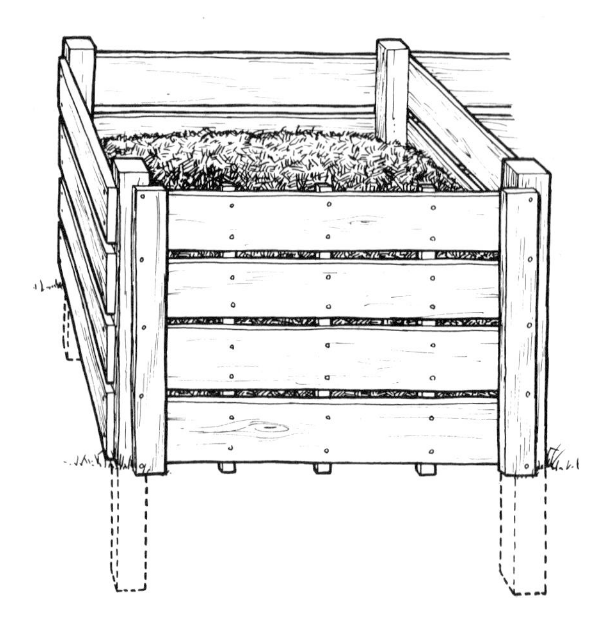

Construction of compost containers

could slide up and down for the opening and closing of the heap. A bank of four such containers was made so that compost was available on a regular basis. While one lot is being used, the second will be ready in about two months. The third follows on while the fourth is in the process of being filled.

Straw and weeds are put in the appropriate container, with a layer of animal manure every few inches to hasten the decomposition process. We try to avoid putting in the roots of perennial weeds such as docks and nettles, as well as twigs and other material which would take a long time to rot down. Inevitably a certain proportion of such things find their way in and, it must be admitted, our compost has been known to include pieces of plastic and baler twine as well as the long-lost potato peeler.

The resulting compost is added to the soil when it has decomposed. Compost ready in late autumn and winter is incorporated during winter digging. If the compost is not completely rotted at this time, it is merely laid on the surface, rather than being dug in. In this way, nitrogen loss from the soil is avoided, and the strawy compost prevents weeds becoming established. During the growing season itself, it is added to the beds, around the vegetable plants, to act as a mulch. This is extremely beneficial for it not only helps to smother weeds, but also conserves moisture at a time when drought could be a problem. Using compost to earth up potatoes is also a good practice.

Planting and Raised Beds

When we had carried out the initial ground preparation for our kitchen garden, we had divided it into several

Compost heaps are essential for maintaining the soil fertility

One of the raised beds with a crop of onions almost ready for harvesting

small plots, rather than having one large, allotment-sized area. The reasoning behind this was that it is easier to work a plot if you have access from both sides without treading and compacting the soil. There is also the psychological aspect which should not be dismissed too lightly. It is surprising the difference in attitude experienced when one is faced with a single large plot and, alternatively, several small ones. Weeding and preparing one plot often seems endless, but with a system of several smaller ones, there is a feeling of satisfaction at completing just one; the whole task seems lessened.

At about the time of initial ground preparation, we heard reports of some American gardeners who were claiming high vegetable yields as a result of using raised beds. These were beds cultivated to a greater depth than normal, which had a great deal of compost incorporated in them so that they were not only deep but were raised above the level of surrounding paths. Crops could be grown closer together because there was less lateral root competition and the leaf canopy of the plants themselves provided a 'living mulch' to resist drought and suppress weeds.

When we researched this particular method of cultivation we found that there was nothing new in it. The mediaeval gardeners of Europe had grown crops in this way, as well as the market gardeners on the outskirts of Paris up to World War One. We even discovered that this form of intensive cultivation had been carried out commercially in Essex at the turn of the century. What had caused its decline was the changeover from horse-drawn transport to the motor car, with a resultant shortage of horse manure. The same process of mechanization brought about a situation where crops were increasingly grown on a field scale, in straight lines, rather than in the traditional 'staggered' planting which was more labour-intensive. The 'staggered' planting meant that there was less space wasted and a greater yield was possible, but it needed smaller beds which would not have their soil compacted by being walked on. In fact, with our small beds, it seemed appropriate to

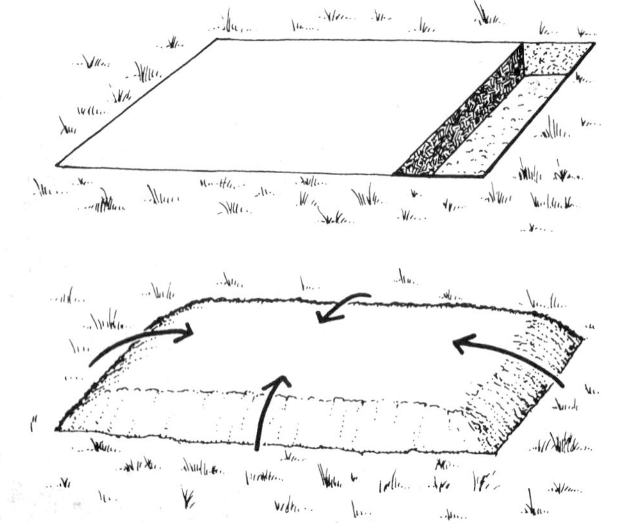

Making a raised bed:
Top *Dig trench and break up sub-soil. Then throw soil from second trench into first, incorporating generous quantities of rotted manure or compost. Continue until completed.*
Bottom *Rake the soil from the sides inwards so that the bed is raised. Do not walk on it to compact soil*

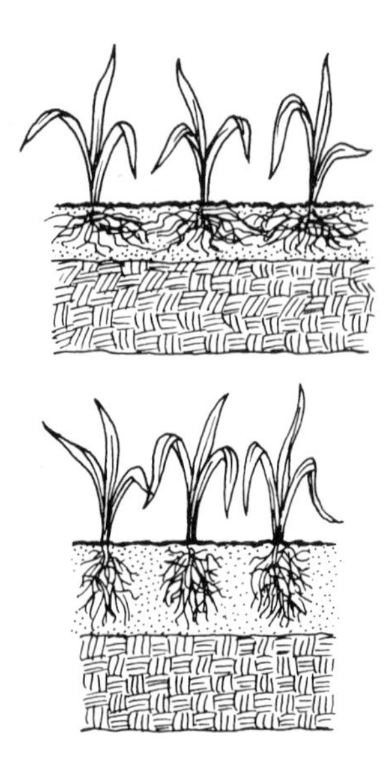

Top *Normal bed. The dotted area indicates the extent of the top soil; the shaded area is sub-soil*
Bottom *Raised bed. In this type of bed crops can be planted closer together: a deeper layer of top soil and manure allows the roots to grow longer, thus preventing lateral competition between them*

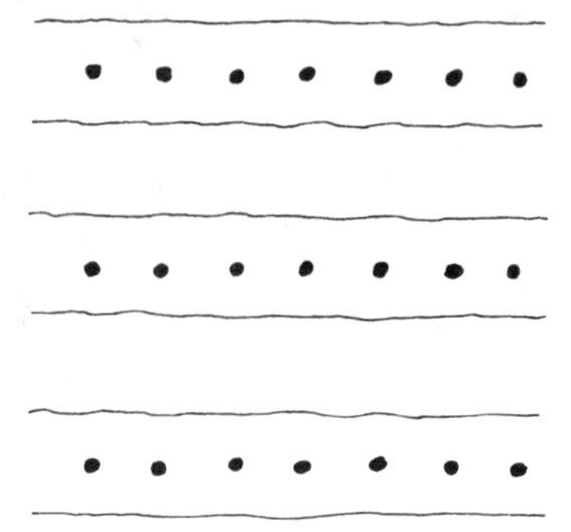

Row cultivation. The spaces left between the rows are not used

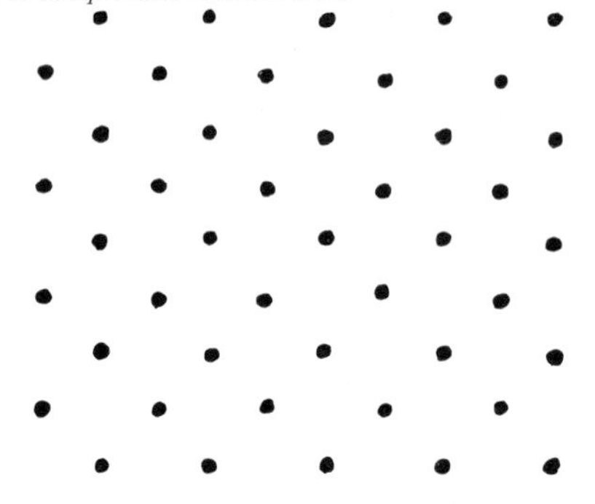

Staggered cultivation. There is no need to walk on the bed, so no space is wasted

give the system a try and we have now grown vegetable crops in this way for seven years. The yields are undoubtedly higher, though nowhere near the four-fold yields claimed by the Californian gardeners; but then, we do live in Essex.

The beds are raised about 15 cm (6 in.) above the path level, but the sides are not wasted for they themselves are used for growing short-rooted crops such as lettuce, radish, alpine strawberries and nasturtiums. Most of our vegetables are inter-cropped, but we do follow a commonsense system of keeping the root crops, brassicas and legumes separate. No crop is ever grown on the same ground two years running, in case soil-borne diseases are transmitted. Apart from this, we interplant as much as possible, so that mono-culture, even on a small scale, is avoided. The practice of mono-culture, where one crop is grown to the exclusion of others, is a comparatively recent idea, and many authorities believe that it, more than any other factor, has lead to an escalation in the problem of plant pests and diseases.

The following is an alphabetical list of the crops we grow in our kitchen garden. Perennial crops, such as some of the herbs, are grown in permanent beds, but annual herbs such as summer savory and sweet basil are interplanted with the vegetables. It is by no means an exhaustive list, and the varieties mentioned are those which we have found satisfactory. It is up to the individual to find which vegetables and varieties suit his particular needs and situation.

Hoeing between crops stops the soil compacting and prevents weed seedlings becoming established

Vegetables

Asparagus A luxury vegetable which will not produce a harvestable crop until at least three years after planting. It likes an open, sunny situation and prefers a soil with a pH of 6·5–7·5. One-year-old corms are the best to plant because they get off to a good start and there is less root disturbance than with older corms. They are planted in trenches 30 cm (12 in.) deep and 45 cm (18 in.) wide which are half-filled with well-rotted manure or compost. Place the corms 30 cm (12 in.) apart on top of this and cover with a 7·5 cm (3 in.) layer of fine soil. Keep the bed well-weeded and resist the temptation to cut any of the stalks. Cut down the foliage when it turns yellow in the autumn and apply a layer of well-rotted manure, then mound up the soil by drawing it with a hoe from

Cut asparagus spears below soil level

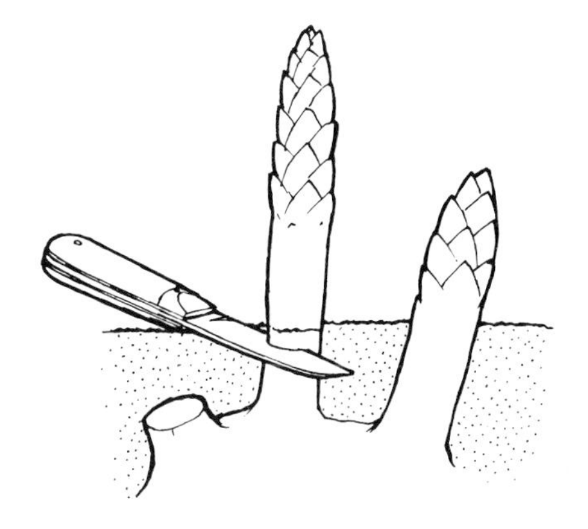

either side. The next year, repeat this. In the third year, cut all the spears, even the thin ones, when they are about 12–15 cm (5–6 in.) high. Cut them just below the soil surface, but restrict the cutting to a period from the end of April to the end of June. After that, allow the subsequent spears to produce foliage and follow the same practice of cutting down foliage and manuring as in previous years. Asparagus is harvested every year after the initial three-year establishing period. The variety we grow is Connovars' Colossal, but growers also speak well of Martha Washington Improved.

Beetroot A popular pickling vegetable which likes a sunny spot and a pH of 6·5–7·5. We sow it direct where it is to grow, ensuring that there is no fresh manure in the bed, for this will have the effect of making it fork (form two separate roots). It can be sown at any time, from April onwards, but if the spring is a cold one, it is usually better to wait until May. It dislikes a cold soil and there is nothing to be gained from early sowings. The other common mistake is to sow too thickly; it is much better to sow in individual stations in a staggered row, with three seeds in each station. The seeds are big enough to make handling easy and after germination the strongest of the three is left to grow on while the others are thinned out. Beetroot needs adequate water, otherwise it will bolt and go to seed. Bird attack may also be a problem, so protective netting may be necessary. We grow the varieties Boltardy and Detroit Crimson Globe.

Broad beans It is a common attitude in the country that only a complete fool could fail to grow broad beans. They are certainly an easy crop to grow, but there can be problems which should not be under-estimated. The main one is mice which eat the large seeds, particularly of early varieties which are sown in November for over-wintering and which are protected by cloches. The other nuisance is blackfly which affects the growing tips, particularly of the spring-sown, later varieties. The best defence against this is to sow as early as possible and then to pinch out the growing tops in late May and early June so that there is less to attract the aphids. Seeds can either be sown direct 15 cm (6 in.) apart in all directions, with a few extra at the end for subsequent transplanting if any fail to germinate, or in pots in the greenhouse or cold frame for planting out later. One of the best varieties for autumn sowing is Aquadulce, while Masterpiece, Green Windsor and Colossal are suitable for the spring.

Broccoli A useful and hardy vegetable which seems to withstand the coldest of winters where many over-wintering cabbages would fail. There are basically two kinds, the purple-sprouting and the white-sprouting. They need a well-manured soil with a pH preference of 6·5–7·5 and can be sown in boxes or a seed-bed between April and May for transplanting to their growing site, from June to July. They are spaced about 30 cm (12 in.) apart and may need bird protection. As with all

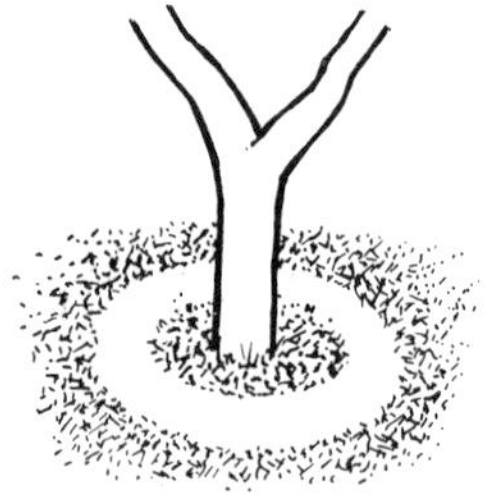
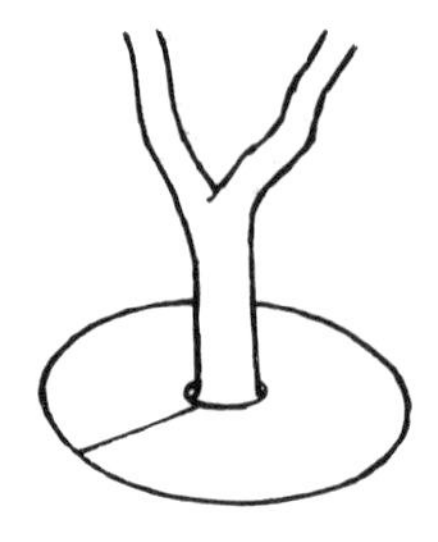

Protecting brassicas:
Left *Circle of lime to guard against clubroot*
Right *Tarred paper collar to deter cabbage root fly*

brassicas, there is a danger of clubroot and we sprinkle an added circle of lime around the stem of each plant as a protection. Cabbage root fly can also be a problem, and collars of tarred paper placed around the stem will give protection. These are now available at most garden suppliers.

Brussels sprouts Sprouts like similar conditions to broccoli and are sown from February onwards, in seed trays, for pricking out and subsequent transplanting. As with all brassicas, we prefer to grow our own from seed because our soil is, at present, free of clubroot and we do not wish to risk introducing the disease by buying in plants which may carry it. Soils which have adequate supplies of lime are less likely to be affected, so the application of lime should never be neglected.

The plants are put in their final places as soon as they have been hardened off and are planted at a distance of 75–90 cm ($2\frac{1}{2}$–3 ft) from each other. They are ready for harvesting as soon as the sprouts are nicely formed, but the flavour is improved after they have been touched by frost, so we always wait for this before picking. Any yellowing leaves at the base should be removed.

Cabbage Spring and summer cabbage need similar conditions to other brassicas (see Broccoli) and are planted with 15–20 cm (6–9 in.) distance between them, depending on the variety and its size. Bird attack can be a nuisance and netting is the best insurance against this. Cabbage white butterflies relish the leaves as a nursery for their caterpillars, and, where chemicals are not used, salt solution spray will deter them. We sow our spring cabbages between July and August, then plant them out in their final positions between September and October. Summer cabbage are sown between February and May and transplanted between April and June in order to have a succession for the kitchen. The varieties we grow are Durham Early and Flower of Spring for harvesting in spring, and Primo and Hispi as summer cabbages.

Cauliflowers The conditions are the same as for other brassicas (see Broccoli). Cauliflowers should be given adequate water when they are transplanted, otherwise they take a long time to establish themselves. Some people even wait until it is pouring with rain before transplanting them, but we have never shown such single-minded determination with our cauliflowers. We sow the summer variety Dominant in the greenhouse in late January and plant out in late March after hardening off. The autumn variety All the Year Round is sown between April and May, for transplanting in June or July. The winter cauliflower Snowball is sown between May and the end of June for transplanting between July and August. These winter cauliflowers need protection: either cloches which should be placed over them in late autumn, or a greenhouse. Some people delay the final planting out until greenhouse tomato plants have been removed at the end of the season, so that the cauliflower plants can take over the vacated bed.

Carrots Our soil is rather heavy for carrots, which like a light, well-drained soil. Initially, we grew only early varieties which were not deep-rooted, but after going over to deep beds, we were able to grow maincrop ones as well. Carrots do not like fresh manure which will make them fork, but completely rotted manure in the soil will not have this effect.

We sow the variety Chantenay Red Cored in March and do not thin them out at all, just pulling them as they are needed for salads or as a delicious cooked vegetable. The maincrop variety Flak is a Dutch strain that we have found to be reliable and full of flavour. It produces long, thick carrots which store well in sand after being lifted. We sow this variety in April to May and thin out to allow adequate room for growth. As carrot fly can be a problem, it is important to carry out this thinning either while it is raining or immediately before or afterwards. The carrot fly is extremely sensitive to the smell of bruised carrot foliage and will be attracted to the area. Rain helps to 'damp down' the smell and makes an attack less likely. The only other precaution we take is to sprinkle salt around the carrot plants to deter the flies. This must be done in dry periods. Carrots and onions are frequently grown together because they are said to have a mutually beneficial effect. This is also our practice, for while many of these claims may be merely 'old wives' tales', there is sometimes a grain of truth in them, and it certainly does no harm to follow them.

Celery This used to be a difficult plant to grow well because there was so much work entailed in ensuring efficient earthing up and blanching of the stalks. The new self-blanching varieties such as Lathom Self Blanching are much easier to grow, and, while they may not be as big or likely to win prizes at shows, are more convenient for those who merely want a few plants for soups and salads.

Celery likes a fertile soil with a pH preferably of 6·5–7·5 and adequate water supplies. It is best sown in boxes between March and April and planted out 15 cm (6 in.) apart from May to June. Once the plants are established, tie the stalks together, just below the leaves. If you neglect to do this, there will still be useable stalks, but they are more successful if this action is taken. They are ready to use from October onwards.

Courgettes All the marrow family, which includes courgettes, vegetable spaghetti, pumpkins and marrows themselves, need the same conditions. Their pH preference is 6 and a fertile, well-manured soil with plenty of water will ensure luxuriant growth. The large seeds can be sown in individual pots in the greenhouse from mid-April onwards and planted out when all danger of frost is past in early June. In mild springs, planting out can take place from mid-May, but it is a good idea to provide cloche protection in case of a sudden frost. Seeds can be sown outside from mid-May onwards.

Courgettes should be picked regularly before they get too big. This will increase the number produced and prolong the picking season. Any that are left will grow into marrows.

The marrow family is rambling and needs plenty of space. They do well on deep beds with an adequate supply of organic matter. We grow them in this way with nasturtiums and borage planted on the sides of the beds. Both of these annuals are said to be good companions for the marrow family, and the nasturtium leaves and borage flowers are good in salads. Nasturtium seeds can also be harvested and pickled as an alternative to capers.

French beans Dwarf or French beans grow well in our garden but we do have to ensure that sowing is not too early in our relatively heavy soil. They dislike the cold, and will usually rot if sown too early. They prefer a pH value of 6·5–7·0 and a fertile soil in a sunny position.

We sow from May onwards, depending on the weather, spacing the seeds 75 mm (3 in.) apart. They need adequate supplies of water, but the soil in between the plants should be forked regularly to stop 'panning' or the formation of a hard crust. They do not need any form of support.

Once ready, the beans should be picked regularly and frequently, otherwise they will get too large and stringy.

Our particular favourites are Masterpiece and Tendergreen.

Leeks We sow Musselborough leeks from late February onwards in seed boxes and plant them out from late April to early May. Holes about 15 cm (6 in.) deep and 10 cm (4 in.) apart are made in the soil with a dibber, and a leek seedling is put in each hole. Trimming the roots and leaves to about two-thirds of their original size before planting helps the leeks to establish themselves more rapidly. The soil is not firmed around the plants, but is left to settle naturally. Once they are watered in, the leeks soon begin to grow. Occasionally during the growing season we add a few inches of compost around

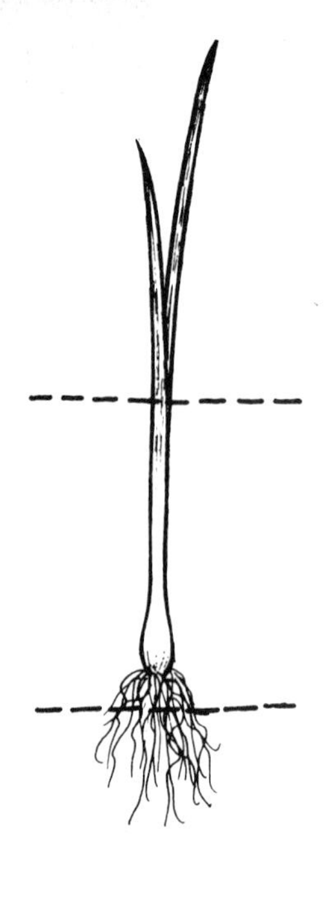

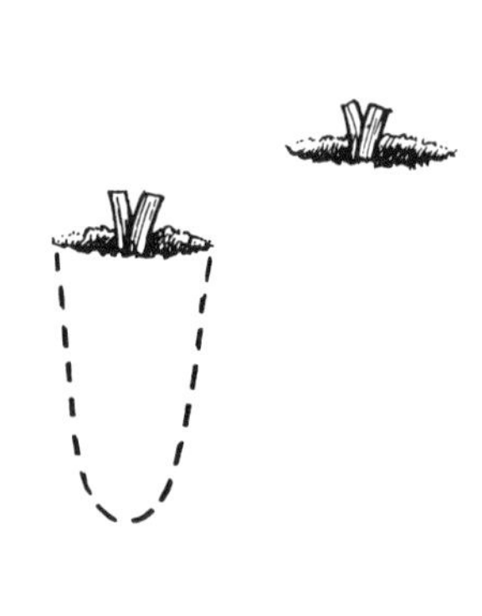

Planting leeks:
Make hole with dibber. Trim leaves and roots of leek seedling.

Drop into a hole 15 cm (6 in.) deep, water in and leave soil to settle naturally. Space plants about 10 cm (4 in.) apart.

them. This provides nutrients in the soil, as well as earthing up the plants to produce a greater length of blanched stem. It is also useful in conserving moisture and smothering weeds. Leeks are the hardiest and longest-lasting of winter and early spring vegetables. Even the 20° of frost experienced in the winter of 1981–82 did not affect them.

Onions We grow several types so that there is never a time during the year when onions are not available. The large keeping onions for storing are planted as 'sets' and not grown from seed because we find that seed onions are more likely to suffer from the onion fly pest. For a very early crop, we plant the variety Unwins Autumn Planting in the autumn, followed by Sturon planted in late February or early March. The small 'set' onions are pushed gently into the soil, so that just the tip shows above the surface. If there is a particularly long 'string' at the top, this is trimmed back in case it attracts birds. If birds are a particular nuisance, it may be wise to net the bed until the onions have established themselves.

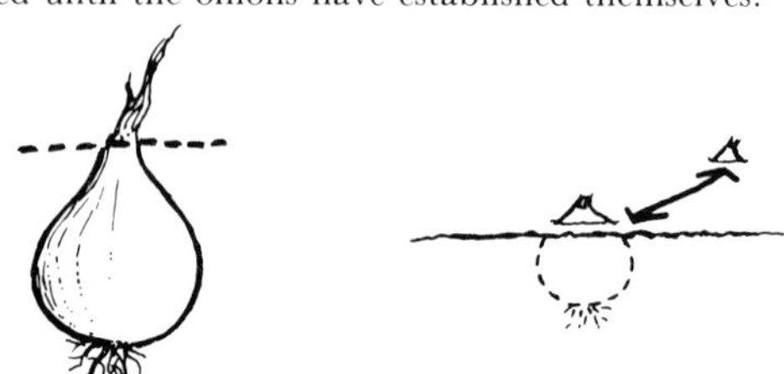

Planting onions and shallots. Plant each onion set individually, about 15 cm (6 in.) apart. Trim off the top

In February or March, we also plant shallots or pickling onions. These are slightly larger than the onion 'sets' and instead of each one growing into one large onion, it produces a number of smaller ones in a cluster.

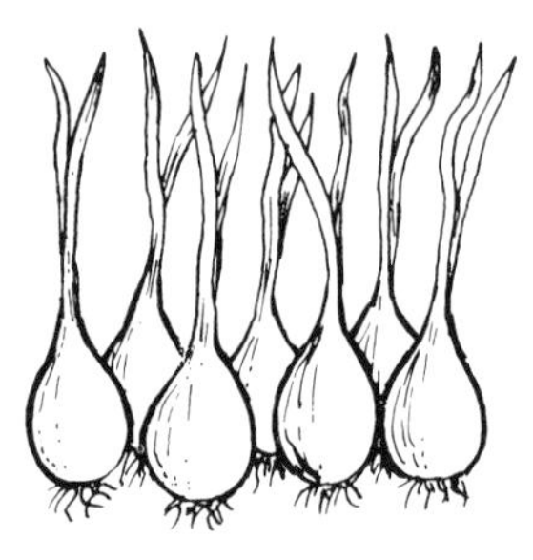

Shallot onions

Both the keeping onions and the shallots are ready for harvesting when the foliage has yellowed and the onions have dried in the sun.

In the depths of winter we rely on the Welsh onion which is grown as a perennial in the herb garden. The individual small onions can be uprooted, or the foliage used instead of chives. Another perennial is the Egyptian or Tree onion which produces a cluster of small onions at the top of the stem.

From spring onwards, we have *chives* in the herb garden. This is a perennial that grows easily from seed, providing green leaves for chopping in salads and stews, but it dies down during the winter. Salad onions, such as White Lisbon, are also sown from spring onwards.

Finally, *garlic* is grown in a similar way to onion sets.

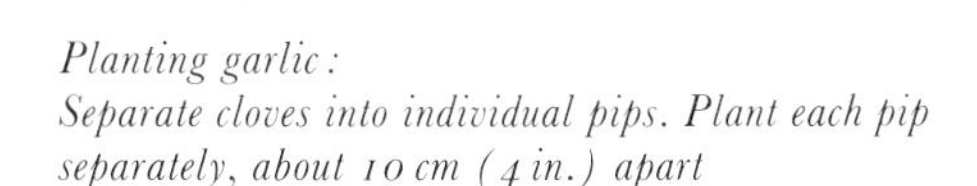

Planting garlic:
Separate cloves into individual pips. Plant each pip separately, about 10 cm (4 in.) apart

A good crop of pickling onions from one small shallot

Individual cloves from shop-bought heads are planted in February. During the season, each clove multiplies and produces the familiar garlic head with its white, papery skin.

All onions need a fertile soil with a pH value of about 6·5.

Parsnips This is another useful winter vegetable which is delicious par-boiled and then roasted in the oven with butter. It is an under-valued vegetable, less frequently grown than it used to be, possibly because it needs a fair amount of room and has a long growing season. It is best sown in February, as soon as the soil is workable, and it likes a pH value of 6·5–7·0. The soil should be fertile without fresh manure, otherwise the roots will fork.

Parsnips are ready for harvesting when the foliage dies down, but they should not be dug up. They are left in the ground and dug as needed; a slight frost improves the taste.

Peas Fresh peas are one of the most delicious of vegetables. Unfortunately, they do take up quite a lot of space and are subject to pest attack in summer. We grow only the early varieties because of this and also because maincrop peas are available cheaply in our particular area. We sow the variety Meteor in autumn and use cloches for over-wintering for an early crop. In spring we sow a row or two of Feltham First, as soon as the weather conditions are suitable, and provide cloche protection. Even when the weather has improved birds can be a nuisance, and netting will still be needed until the plants have made substantial growth. Peas like a well-manured soil with plenty of water, and they are best grown in rows so that supports such as stakes and wire can be provided for them. Their preferred pH value is 6–7 and, once the pods begin to fill out, they should be picked regularly.

Potatoes As with peas, we concentrate on early varieties, for maincrop potatoes are available cheaply in our area and the early ones are not as likely to succumb to blight. Potatoes like a pH value of 5–6, so they should be planted in a bed which has not been limed. Lime in excess of their needs will cause 'scabbing' of the skins. We plant potatoes from the end of March onwards, depending on the weather, and provide them with plenty of well-rotted manure. They are planted in rows 60 cm (24 in.) apart, with 20 cm (9 in.) between each plant. The trenches are dug 30 cm (12 in.) deep, a few inches of well-rotted manure are placed in the bottom and the seed potatoes are put on top. They are then covered up and, as the first stalks appear, soil is drawn up around them. We keep a bale of straw to hand nearby, so that if there is a frost warning straw can be placed on top of the emerging plants to protect them.

Potatoes should be set out to 'chit' or sprout before planting

Comfrey leaves placed between the potato plants act as a mulch and provide instant fertilizer

Once this risk has disappeared, the plants are kept earthed up, so that only the leaves show above the surface. The best way we have found of doing this is to add compost on either side of the rows and draw it up around the plants. It provides nutrients for the plants, as well as a moisture-retaining and weed-suppressing mulch. We also cut long stems of comfrey and place them between the rows. As comfrey rots down, which it does quickly, it releases nutrients into the soil and provides an 'instant' fertilizer.

The varieties we grow are Pentland Javelin and Epicure.

Runner beans A vegetable which is most economical of space, and, as long as adequate supports are provided for it, will produce a good crop of beans for day-to-day use, or for the freezer. The large seeds are sown direct in the soil from late April to the end of May, or they may be sown in pots for subsequent transplanting.

We grow our runner beans in a double row, but many people, particularly those with restricted space, grow them up a wigwam structure of poles spaced out at the bottom and tied at the top. They need plenty of well-rotted compost or manure, and it is our practice to prepare trenches for them early in the year. Adequate water is also important for any degree of drought will cause a premature flower drop.

The varieties Streamline and Kelvedon Marvel have both proved reliable in our kitchen garden. The annual herb Summer Savory is a good companion plant for the beans.

Spinach This is a vegetable which is superb when cooked with a few teaspoonfuls of water only, then put in a blender with butter, salt and black peper. It is a good source of fibre in the diet.

Spinach prefers a pH value of 6·5–7·5 and will do well in a slightly shaded area, such as between taller crops. It

Runner beans need a good support that will stand up to the wind

Removing side shoots from a tomato plant. Pinch out side shoots as indicated by arrows

can be sown, in succession, from late March onwards and an early autumn sowing will provide an early spring crop.

Tomatoes We grow tomatoes as a protected crop in a greenhouse and conservatory. Although certain varieties will produce outdoor crops, they are dependent upon the weather and in a cold wet summer will not ripen satisfactorily. This does not mean that tomatoes are not worth growing outdoors, merely that more reliable crops are possible in protected situations. It is now possible to buy purpose-made tomato cloches which are tall enough to house the plants satisfactorily. There is a wide range of varieties available, but our particular favourite is Pixie which can be grown either in the normal way – pinching out the laterals and training it up supports, or in pots as a bush plant. It has a superb flavour and produces medium-sized fruits.

We sow seeds in pots in March for planting out in late April in the greenhouse. As side-shoots are produced, they are pinched out so that the plants are trained up supports. These can be canes to which the stems are tied at different heights, or simply strings.

Other salad crops Lettuce is deservedly the most popular salad vegetable and has a place in every garden. There are many different varieties, but we find that All the Year Round is precisely what its name implies and can be sown in spring, summer or autumn. For over-wintering under cloches or in a cold frame or greenhouse, the Dutch varieties Valdor and Kwick are reliable, while for really confined places, Tom Thumb is perhaps the most appropriate. As we have a number of raised beds, we grow lettuces on the sides of these, and here the best variety is undoubtedly Salad Bowl. This has finely divided, attractive leaves which can be picked in great handfuls, without uprooting the plant. The seed is sprinkled thinly on the soil and then left to grow; there is no need to thin out the plants. The result is a solid bank of leaves which does not allow weeds to grow through and which conserves moisture for the rest of the bed. Picking can continue right through the growing season, without danger of the plants bolting. In my view, it is the best and most economic of lettuce varieties. Lamb's lettuce or Corn Salad, sown in August, will provide winter salads.

Radishes These can be grown anywhere where there is a spare patch of ground; it is merely a matter of sowing a pinch of the seed every week, so that not all the plants are ready at the same time. They are quick-growing and must be picked before they bolt and become woody. Even if they have been left too long, they need not go to waste as far as livestock are concerned, for both rabbits and goats will welcome them in their diet. The varieties we grow are French Breakfast and White Icicle. The latter is pure white and forms a long, tapering root, making a novel addition to salads. China Rose is a variety for sowing in mid-summer or early autumn for winter salads. It produces thick radishes, up to 5 cm (2 in.), and provides a welcome taste of summer when winter arrives.

Herbs

No kitchen garden is complete without at least a small patch of herbs. The culinary herbs enhance a wide range of dishes such as salads, stews, roasts and puddings. Minor complaints such as colds and coughs can often be aided by the use of soothing herbal drinks and livestock appreciate small amounts of herbs in their diet. Many herbs are excellent bee plants and they can also be used in the manufacture of scented articles such as lavender bags and *pot pourri*.

Our herb garden was created from what was once paddock and is to one side of the kitchen garden. It backs onto the south-facing wall of the garage so that conditions are sheltered and sunny: an ideal situation for herbs. Most of the perennial or permanent herbs are planted here, with stone flags as stepping stones in between them. Taller ones such as tarragon are planted at the back of the border, while low-growers such as marjoram and thyme are at the front.

Many perennial herbs can be grown from seed sown in pots at any time from March to July. The young plants are then planted in their permanent positions in the autumn. Annuals are sown from March onwards in pots for transplanting in May or June, or sown directly where they are to grow from April onwards. Some crops such as parsley are not annuals, but are best treated as such. A new sowing each spring gives a better and more lush crop.

The range of herbs is wide and only the largest of gardens would have room for them all. The following is a list of the ones that we find useful to grow; but, again, it should be emphasized that this is a personal choice.

Balm The leaves of this herb are delicately lemon-scented, giving it its alternative name of Lemon Balm. It is a tall-growing perennial, easy to grow from seed or from cuttings. It is relished by goats and rabbits as an addition to their browsing diet and the young leaves are pleasant in salads. Perhaps their best attribute is as a tea, when steeped in boiling water. A bunch of leaves can also be added to ordinary tea leaves in the teapot; if the tea is a poor quality one, it will be transformed into a delicate connoisseur's brew by the balm.

Basil This is a rather tender annual which only does well outside in hot sunny summers. It grows well in pots and we also find it excellent grown in the greenhouse in between the tomatoes. Whenever we pick tomatoes we bring in some basil leaves at the same time and chop up the delicate aromatic leaves for sprinkling on the sliced tomatoes. It can be sown from spring onwards.

Borage Unlike basil, borage is a hardy annual and once you have it in your garden, it seeds itself regularly, appearing even after the hardest of winters. It frequently appears in between other crops and can be left, for it will do them no harm. The rather prickly leaves have a cucumber flavour while the bright-blue star-like flowers are said to 'gladden the heart'. They are certainly attractive when sprinkled on salads or floating in cool summer drinks, but the hairy calyx of sepals must be removed first.

Caraway This is grown for its seeds which are excellent when cooked with cabbage, added to fruit cakes or sprinkled on home-made bread rolls. It is a rather drab-looking plant which has the advantage of doing well in semi-shaded positions. It is sown where it is to flower at any time from March onwards, depending on weather conditions. The seeds are gathered once they are dry; it is best to pick the whole heads and hang them upside down with a paper bag around them to catch the seeds as they fall out.

Comfrey This is grown mainly as a fodder plant for the goats, rabbits and chickens and as home-produced fertilizer for the garden. It is a deep-rooted perennial and does best when grown in its own separate bed. It grows quite high and, as a fodder crop or fodder hay, it is best cut before the flowers form. You can cut it down to the ground, three times a year, and it will rapidly grow again.

If the cut stems and leaves are laid in between potato rows, they provide a good natural fertilizer and also serve as a mulch. When allowed to flower, the blooms are much appreciated by foraging bees.

Coriander This is similar to caraway and is grown for its seeds which are useful in curries or in home-made pickles. It is sown in semi-shaded areas from spring onwards, and the seeds are harvested in a similar way to caraway.

Dill The feathery leaves of dill are excellent with home-grown gherkins and a sprig put in the pickling vinegar will greatly enhance them. Dill is also one of the main ingredients of 'gripe-water', which explains why it has traditionally been a favourite remedy for colic in infants. Boiling water poured onto some of the leaves is left to steep for a few moments then strained. A sipped teaspoonful of the cooled liquid is beneficial for many digestive upsets.

Perhaps one of the more unusual uses of dill is as a camouflage. One year I sowed the dill in rows, with summer cabbage sowed in between a few weeks later. The feathery plumes of the dill provided excellent cover

The herb garden at Broad Leys

for the young cabbages which completely escaped the attention of ravenous pigeons that year.

Horseradish Connoisseurs of British roast beef would say that it is incomplete without horseradish sauce. As we produce our own beef we could hardly not grow horseradish. It likes plenty of moisture and is therefore happy in relatively heavy soils. It is rather invasive and unless you want it spreading through other plants, it is best grown on its own in a separate bed. It is the roots which are used, and these are dug up, washed well and grated. It will bring tears to your eyes as you grate, but once turned into sauce will be worth the tribulation.

Hyssop One of the perennial herbs, hyssop has beautiful dark blue flowers, much loved by bees. It is a small shrub and the leaves, when stripped off a picked stem and chopped, add flavour to meat roasts. It is one of the so-called 'purging' herbs and has a strong flavour, so use it with discretion.

Lavender This is so well known that it needs little description. The flower stems are cut when the flowers are at their best, and hung to dry in a cool, airy place. Again, a paper or muslin bag will catch any flower heads that fall. Once dry, the heads are stripped off the stems and used to make lavender bags or *pot-pourri*. Lavender benefits from an annual clipping after flowering, otherwise it will become woody and straggly.

Marigold No herb garden is complete without some pot marigolds, so called because it was traditionally regarded as one of the culinary 'pot' herbs, not, as some think, because it was always grown in pots. The petals are excellent and attractive strewn on salads and they give stews a piquant flavour. It is important to remember that the marigold referred to here is *Calendula officinalis*, not the African or French marigolds which are frequently grown as half-hardy annuals for the flower border.

Marjoram Another of the popular 'pot' herbs, marjoram is good in stews, finely chopped in salads, or sprinkled on meat and fish before cooking. It can be grown from seed or is easily propagated from cuttings. The flowers are attractive to bees and are also suitable for *pot-pourri*. It is relatively low-growing and quickly forms a dense cushion at the front of the herb border.

Mint There are nearly 30 different mints, including Eau-de-Cologne, Ginger and Variegated, as well as the more common Apple and Spearmints. They are perennials and all love damp, shady situations. They are extremely invasive, so care must be exercised in selecting a position. We grow the large and hairy-leaved Apple mint as well as the smaller Spearmint, because a blend of these two varieties was always considered to be the best combination for mint sauce to accompany Welsh lamb.

Nasturtium Some people may consider this to be more appropriate in the flower garden than in the kitchen or herb garden. It is certainly an eye-catching plant with its rambling growth and bright orange flowers. We find it useful to grow on the sides of raised beds, at the corners of vegetable plots and anywhere where there happens to be a space. The large seeds are just pushed under the soil, from March onwards, and will grow without any difficulty. The dwarf varieties are better in confined spaces, or at the front of borders, but there are also climbing varieties suitable for hiding an old fence or other eyesores. The young leaves are nice in salads, while the pickled seeds provide an alternative to capers.

Parsley Some people maintain that parsley is difficult to grow from seed, but we have never found it to be so. The problem may have arisen because it was once the practice to sow it directly outside in the spring, and a cold soil will deter virtually anything from germinating. We find that the best policy is to sow it in seed trays in the conservatory, then transplant the seedlings outside when the weather is relatively warm. It needs shade and plenty of water and is excellent in stews, salads and as a garnish. It is a good source of iron and other minerals, and I feed it regularly, in small amounts, to the rabbits.

Rosemary Another of the shrub perennials which is rather tender and may need protection during the winter. In the severe winter of 1981–82 virtually everyone in our area of Essex lost their rosemary shrubs and there was a period when replacement plants were virtually unobtainable. Fortunately they are easy to propagate from cuttings, and it is a wise precaution to take some cuttings from July onwards and keep them in pots in a protected place over the winter in case the mother plant is killed by frost.

The chopped leaves are excellent sprinkled on roast meats, particularly beef and lamb, but use it sparingly for it has a pungent flavour.

Sage A culinary herb which has never lost its appeal, and has always been an important part of the Christmas tradition, in the form of sage and onion stuffing. If the leaves are to be picked for drying, it should be done before flowering takes place. After flowering, the plant should be clipped back to prevent it becoming straggly.

There are several varieties of sage, the normal culinary type, the variegated one and a purple-leaved type which is often grown by beekeepers.

Savory A herb which has been grown as an alternative to pepper. Its hot-tasting leaves are also good in curries or in meat sauces. There are two types, the Annual or Summer Savory which is sown in the spring and early summer, and the perennial Winter Savory, which forms a small shrub.

Tarragon The variety to choose is French Tarragon which has a delicate flavour when used in savoury sauces. A sprig placed in the bottle will also considerably enhance a plain wine vinegar. Russian Tarragon is a much more coarse-flavoured variety and should be avoided by discriminating cooks.

Thyme There are many different varieties of thyme, including dwarf or prostrate ones to plant in between flagstones. It is an excellent bee and butterfly plant and forms a small shrub, easily propagated from cuttings. The leaves, dried or fresh, are an excellent accompaniment to roast meat, particularly lamb. It is also relatively easy to grow from seed.

3 Fruit Growing and Protected Cultivation

Fruit Growing

A relatively cool and damp climate is almost ideal for certain types of fruit, notably apples and pears, and soft fruit such as raspberries and blackcurrants. Britain has a long and honourable tradition of selective breeding and hybridizing, having produced thousands of different varieties which have found their way all over the world. These varieties were bred for a range of different soil types and to provide fruit over as long a season as possible. There are, for example, over 6,000 different varieties of apple, encompassing early, mid-season and late croppers which provide picking fruit and stored apples throughout the year. A combination of market forces and the effect of E.E.C. regulations has resulted in a situation where only 300 varieties are available commercially and most nurserymen stock only a few approved varieties. Even fewer firms now list more than 20 varieties in their catalogues, and the wealth of British apple varieties is in danger from commercial pressures that would have us all eating varieties such as the extremely marketable, but relatively tasteless Golden Delicious. Lawrence D. Hills's booklet *The Fruit Finder* (see Bibliography) is an excellent guide to the choice of varieties, listing those which have a good flavour, vitamin content and hardiness, rather than those which have only the marketable qualities of size and colour. He appeals to all those who plant orchards to draw a map of the site, with the names of the fruit trees marked on it, then to nail the map to an attic roof beam. His fear is that in future, it may no longer be possible even to identify existing trees.

Our own plan was to map out and identify the existing fruit trees on the site, then to plant new ones in order to fill in such seasonal gaps as there might be. We wanted to have fruit available for as long a period as possible, but our choice of varieties would be influenced by the following factors: flavour, vitamin C content, hardiness and cropping ratio.

Top Fruit

The term 'top fruit' is applied to fruit trees such as apples, pears, plums or damsons, while 'soft fruit' is that which comes from herbaceous plants such as strawberries or shrubs such as gooseberries. In a small garden it may be possible to have only one or two trees, unless the smaller cordon trees are planted against a fence.

Where more land is available, an orchard is appropriate. In this situation, the fruit trees can be planted to a convenient pattern, allowing plenty of space for grass-cutting machinery. If the orchard is fenced off, it is possible to graze geese or sheep under the trees, so that the land has two functions rather than one. It is worth mentioning, however, that a degree of protection may be necessary, depending on the stock. Adult geese may damage young trees with low branches when there is fruit on them. They reach up for the fruit and may snap the branches as they do so. Sheep may eat the bark, particularly in winter, and it may be necessary to place wire mesh guards around the trunks.

Tackling a Neglected Orchard

At Broad Leys we inherited a small orchard which had been neglected for a number of years. The trees were overgrown with a tangled mass of growth and, although producing delicious apples, were badly in need of pruning.

The art of pruning is in cutting the right branches at the appropriate time and on a regular basis. Once a tree has been neglected for a number of years, it is virtually impossible to bring it back to the ideal condition. The danger is that if too much growth is removed, you will kill the tree, either as a result of a massive setback in its general growth or because infection sets into a wound. The former consideration demands that not too much wood is removed at any one time; the ideal is to open up the centre of the tree and cut out any crossing or badly situated branches. This must be done in the winter when the trees are dormant. To prevent infection, all cuts should be clean and on a slant so that rain water is shed. If the wounds are painted with a proprietary sealing compound, there is no danger of disease or fungus becoming established. Once the trees have been brought back into reasonable condition, the normal regular pruning can take place each year.

Buying Fruit Trees

There are several forms in which trees are available to the purchaser. The *standard* is the largest and has a trunk of approximately 1·8 m (6 ft). It is grafted onto a

Matthew Thear harvests fruit from the oldest apple tree at Broad Leys – an old Essex variety D'Arcy Spice

vigorous root stock, takes a considerable amount of room and does not begin to fruit for a long time. It is suitable for a long-term orchard, but not for those anxious to have crops in a short time or whose growing space is restricted. The *half-standard* is shorter and takes less space while the *bush type* is smaller still and is probably the best and most easily managed for smaller gardens. In really confined situations, it is best to concentrate on *cordons* which are trained against a fence.

In Britain fruit trees are bought ready-grafted onto classified Malling root stocks and the type of stock used has an important bearing on the resultant tree. A vigorous stock such as Malling 2 produces large standard trees while a stock such as Malling 26 is less vigorous and produces a dwarf tree which will crop only two to three years after planting. In the USA, the United States Department of Agriculture will advise on sources of disease-free stock.

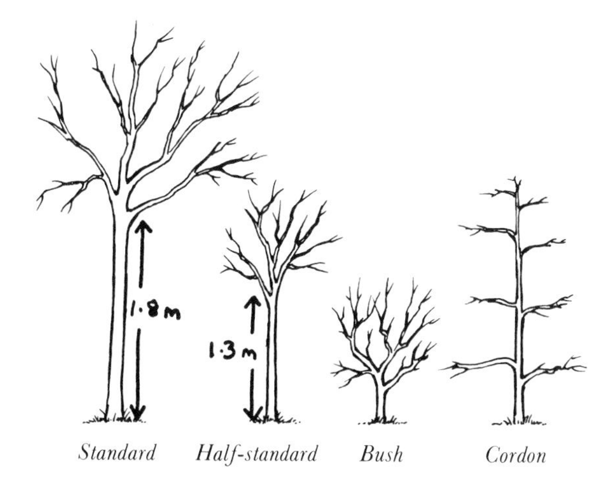

Types of fruit tree

Before buying trees it is important to take these factors into consideration and to choose the right type of tree for the space available. A specialist fruit nursery will be only too pleased to give advice on this question. As far as varieties are concerned, the ideal situation is to have early, mid-season and late varieties of both dessert and cooking fruit.

Apples The Bramley is deservedly the most popular of cooking apples. It does, however, have a spreading growth, making it more suitable for orchards. A variety such as Arthur Turner, with its relatively upright growth, would be better in smaller gardens. It is a comparatively early variety, however, and will not keep for long beyond the autumn. Grenadier is also a good early cooker, but does not store. Bramley should last until February if stored carefully. For longer storage a variety such as Woolbrook Russet, Encore, or Edward VII would be needed.

As far as dessert apples are concerned, one of the earliest is George Cave. Good mid-season varieties are Merton Charm, James Grieve and St Edmund's Pippin, while the famous Cox's Orange Pippin follows afterwards. Really late varieties for storage are Sturmer Pippin and Tydeman's Late Orange.

After taming our orchard at Broad Leys, we discovered that we had several Bramley cooking apple trees, a very early George Cave dessert, an Egremont Russett mid-season dessert and an extremely late old Essex dessert apple, D'Arcy Spice, which stores extremely well. We decided to plant Lord Derby, a particular favourite with us because of its superb crisp flavour. Strictly speaking it is classed as a cooking apple, but we like its sharp tang and it makes a good dual-purpose tree.

Pears There were no pear trees in our orchard so we planted a half-standard Conference and Doyenne du Comice. In addition we planted six cordons, a Conference, Williams' Bon Chretien, Winter Nelis, Louise Bonne of Jersey, Beurre Hardy and Merton Pride. As many pear trees, as well as apple trees, are not self-fertile, it is important to ensure that several are planted, with the flowering periods coinciding so that there is adequate cross-pollination.

Plums Plums grow well in our particular soil, and we planted the Victoria which must be everyone's favourite. It is a dessert and cooking plum and bottles well. It is also self-fertile and a regular cropper. Czar is a readily obtainable deep purple plum which is a dual-purpose fruit for cooking and dessert. It is also self-fertile. Damsons are popular, not only for bottling, freezing and eating, but also for making damson wine. As we did not have any in our orchard we planted the self-fertile variety Shropshire Damson.

Planting Trees

The best time to plant fruit trees is between December and February when they are dormant, as long as the ground is free of frost. The hole should be big enough to allow the roots to be spread out comfortably, and a supporting stake will need to be placed in the hole before the soil is replaced. Attempting to hammer it in afterwards may damage the roots. A little peat is placed around the roots, together with a sprinkling of bonemeal, but too much peat or organic manure is avoided otherwise it will make the tree unstable and liable to be blown over by the winds. It is much better to apply a mulch of organic manure on top of the soil, after planting. The tree should be planted to the depth of the original soil mark on the stem. The grafted stock should be above ground otherwise it will form roots above the join.

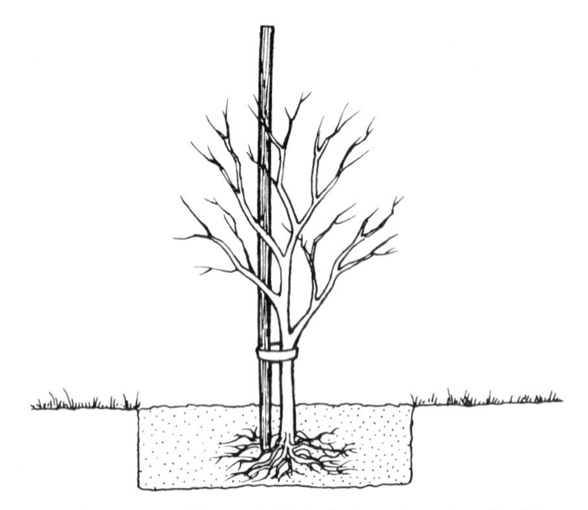

Planting a tree. The grafted join is just above the soil. Tie the tree to a supporting stake

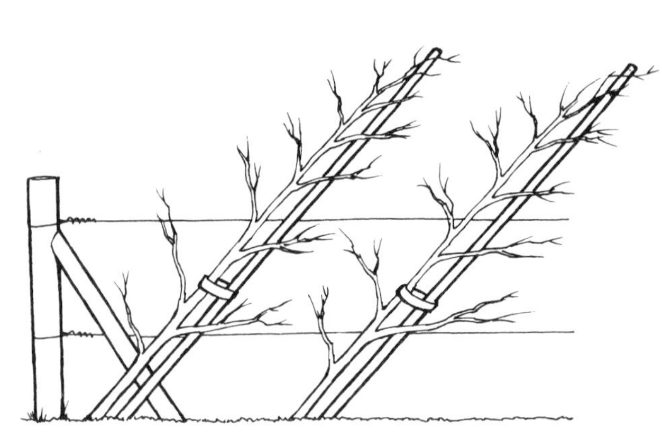

Planting cordons. The cordons are tied to canes which are fixed to the horizontal wires. The straining post is shown on the left

A newly-planted tree needs a supporting stake until it is established

The distance between trees will depend on the type, but, as a general rule, the following applies:

Standards 6–9 m (20–30 ft)
Half-standards 4·5–6 m (15–20 ft)
Bush 3·5–4·5 m (12–15 ft)
Cordon 75–90 cm ($2\frac{1}{2}$–3 ft)

Cordons are grown at an angle of approximately 45°; in other words, they are planted 75–90 cm ($2\frac{1}{2}$–3 ft) apart and then tied into supporting wires so that they all lean to one side. This makes effective use of a confined space, and regular pruning keeps the trees from encroaching upon each other.

Soft Fruit

Soft fruit bushes deserve a place on every smallholding. Like most perennial plants, they require comparatively little time spent on them. As long as the bed in which they grow is kept heavily mulched, there is no need to dig every year. Our practice is to add a thick layer of strawy manure from the goat pens, so that the soil is completely covered in between the blackcurrant and gooseberry bushes. As it is a surface mulch, it is not necessary for it to be completely rotted down. Nitrogen depletion only takes place to a marked effect when the manure is actually incorporated into the soil. It is necessary, however, to ensure that unrotted manure does not touch the stem of the bush, in case 'scorch' damage results.

Blackcurrants and gooseberries are both greedy when it comes to potash, and any deficiency results in poor fruit formation. Wood ash is an excellent source of potash and woodstoves produce this commodity in abundance. We sprinkle the ash around the blackcurrants and gooseberries on top of the mulch and let the rain wash it in. Again, it is important not to let it come into direct contact with the plant stem. It is also important to make a distinction between wood ash and solid fuel ash. The latter is not suitable, although it can be added to the compost heap and left for a few months to become incorporated into general compost. During this period, any mineral wastes are leached out and will not harm the plants.

Blackcurrants Early varieties of blackcurrants are Boskoop Giant, Laxton's Giant and Mendip Cross. Good mid-season ones are Wellington XXX which has a superb flavour, Seabrook's Black which is resistant to big-bud disease and Ben Lomond. Baldwin is one of the late varieties and also has the highest vitamin C content of all the varieties, so is worth growing for this alone. Another late variety is Westwick Choice which is a favourite for bottling. Amos Black and Jet are later still, although their yields are often not as high as some earlier varieties.

Gooseberries Careless and Leveller are the two most easily available varieties of gooseberries. The former is a culinary type, with fruit which bottles well, while the latter is a dessert variety which can also be used for bottling and freezing.

At Broad Leys there were two obviously old and neglected gooseberry bushes which we were not able to identify. These were pruned and manured and brought back into reasonable cultivation. In addition, we planted some new bushes – Careless, Leveller, May Duke and New Giant. These settled in well and the following spring made good growth. Unfortunately, the geese discovered a weak patch in the fence, pushed their way into the garden and devoured all the new green shoots. I have often wondered whether it was coincidence that they concentrated on these bushes or whether the name gooseberry is an indication of the goose's liking for the plant. Whatever the answer, there were no gooseberries on those bushes for two years, but they have since cropped regularly and well.

Raspberries Raspberry canes can be planted in winter, cut back hard and the new growth will be vigorous. The canes are tied onto wires or other supports as they grow, ensuring that they are not damaged and that the fruit is easily picked. The supporting structure needs to be well anchored and braced in case of wind damage. The simplest construction is made from stakes hammered into the ground at each end of the row, with several wires strung taut between them.

As with all fruit, it pays to acquire disease-free, certified stock, and regular bulletins on sources of such stock are published by the Ministry of Agriculture and Fisheries. Varieties that provide raspberries in succession from early summer to autumn are: Malling Promise, Glen Cova, Delight, Malling Jewel, Zeva and Heritage.

Strawberries Strawberries, in my experience, are far more difficult to grow well than any of the other soft fruits. Again, it is essential to buy certified, disease-free stock and to plant them in the autumn in well cultivated and fertile soil. Royal Sovereign is an old, fine flavoured variety while Cambridge Favourite, a vigorous and reliable grower, follows afterwards. Gorella and Red Gauntlet are mid-season croppers while Talisman provides late fruits.

All soft fruits are liable to bird attack, but strawberries seem to be particularly sought after. Netting is the only solution, but it must be securely pegged down, for blackbirds, in particular, are amazingly adept at finding a way in. Early crops can be obtained by using cloches in the early part of the year. It is important to remove them or open them up, to allow adequate ventilation and pollination to take place. Once the fruit forms, straw should be placed underneath, so that the berries are protected from the soil. Slugs may be a problem and controlling them can be difficult. There are several organic remedies now on the market which do not introduce chemicals.

One of our main problems at Broad Leys has been the depredations of pigeons. Fruit bushes could be adequately protected, but strawberries were so difficult to protect that we gave up outside crops and concentrated on using one side of a polythene greenhouse instead. This means that we have early crops of dessert strawberries. For jam-making and bottling, we rely on 'pick-your-own' strawberry growers who are plentiful in our area.

It is important to remove the runners of strawberries, unless new plants are to be taken from them, otherwise the yield of fruit per plant will decrease. After four years, a new bed should be started.

Rhubarb Rhubarb is such an easy plant to grow that it deserves a place in every garden. We have turned one of our beds into a perennial rhubarb bed, and the only treatment it ever receives is the application of a thick layer of manure every autumn. It is never dug and weeds are kept at bay, not only by the spread of the plants' large leaves, but also by mulching with grass cuttings during the summer. We planted Timperley Early and Champagne which are both good varieties for forcing early in the year. Forcing is simply a matter of placing a container over the shoots when they first appear through the ground, and excluding light from them so that they grow long stems in a short time. A bucket or wooden box is suitable, and if straw is placed around the container, it provides added insulation against possible frost damage. Our rhubarb forcer is a cast-iron pot which came from our old, dismantled Aga cooker.

Protected Cultivation

Early, late or generally out-of-season crops are possible if a protected environment is provided. The cheapest and

Rhubarb is an easy, perennial crop and can be 'forced' to produce early sticks by covering with a pot

simplest method is to use cloches. These are either glass or plastic and there are advantages and disadvantages with both materials. Glass is more expensive but, with care, lasts longer than plastic. It lets in light more efficiently, but is more of a hazard, particularly where there are young children in the family. The type used is a matter of individual preference but whatever the choice, the basic principles of use are the same.

Where crops such as lettuce or broad beans are to be over-wintered, the cloches should be in place by late October to make sure that adequate protection is available against a sudden frost. A careful watch needs to be maintained to ensure that there is enough ventilation and that the crops are not being eaten by slugs. If the cloches are being used to start off early crops in the spring, it is important to place the cloches in position for about two weeks before the plants are sown or planted. The effect of this is to warm up the soil beforehand, so that the plants get off to a good start'.

A greenhouse, whether plastic or glass, is useful, not only for over-wintering plants and having early crops, but also ensures that, in a poor summer, good crops of tomatoes, sweet peppers and melons are still possible. The great disadvantage of a greenhouse for the small grower is the cost of heating it. To be fully used it needs to be heated, and the cost of fuel and energy is such that the average smallholder or householder with a heated greenhouse is probably losing money.

At Broad Leys, we went into this question in considerable detail and decided that a glasshouse was not economic. A polythene tunnel was, but only if it was unheated. We therefore decided to buy a tunnel measuring 9 × 2·4 m (30 × 8 ft) and this is used for growing early strawberries, maincrop tomatoes, sweet peppers, cucumbers and melons. Because it is unheated, it means that the tender crops cannot be planted there much before April. We still needed somewhere where early sowings could be made and which could be heated on a cost-effective basis. We decided that the only possibility was a conservatory, attached to the house, which would utilize the main central heating system, and, as described earlier in the book, we built a conservatory on part of the verandah on the south-facing side of the house. It has a double radiator connected to the house central heating system and a roof made of tough crystal-clear corrugated plastic. Thermal blocks were used to brick in the area between existing brick pillars on the verandah. Above these, two 2·1 × 1 m (7 × 3 ft 6 in.) windows were inserted and a half-glazed door was placed to one side of these. Two strip lights were fitted, as well as an electric point on the wall. The completed conservatory measures 6·6 × 2·1 m (22 × 7 ft) and utilizes the existing verandah flagstones for the flooring.

The advantages of such a conservatory are many. It is an integral part of the house. It provides an extra room for various activities and makes gardening a practical proposition, regardless of the weather. The cost of heating is reduced by the fact that it is connected to the central heating system. What is lost in heat output for the conservatory itself is gained in overall insulation of that side of the house.

In winter, further insulation can make a great reduction in the amount of heat lost through a conservatory roof and windows. There are many ways to achieve this and, in recent years, a wide range of insulating materials and fittings has become available. One of the most effective is a form of 'bubble' plastic which incorporates air and so has good insulating properties. Purpose-made attachment clips are sold with it.

In our conservatory, we use ordinary horticultural polythene cut to size and fitted with paper staples. It is simply a matter of cutting small squares of cardboard, placing them over the polythene where it is to be attached and stapling through to the wooden roof beams or window frames. The cardboard prevents tearing of the polythene. It is a cheap and simple method of insulation, but is amazingly effective. It is put up in late

Small, ready-to-assemble conservatories are available (Banbury Homes and Gardens Ltd)

Below *The home-made conservatory built at Broad Leys for out-of-season production of food crops*

October and left until April when it is removed and stored ready for re-use the following winter. The real test of its effectiveness came in January 1982, the worst winter in living memory in Britain. At that time we went to see relatives in Australia for three weeks, leaving the house central system on its minimum setting, just enough to prevent the pipes freezing. When we returned the plants in the conservatory were thriving, although outside temperatures had reached −20 °C (−1 °F) in our part of Essex.

A conservatory such as this makes over-wintering of precious plants quite easy, but in order to provide maximum benefit, it needs actively to extend the growing season as well. One of the best ways of achieving this is to grow your own plants from seed. Not only is it much cheaper to do this, but there is no risk of introducing infected plants onto your land. Mention has already been made of the risk of introducing the disease clubroot from bought-in brassica plants.

Seed trays or pots will be required, together with a good quality seed compost. Small peat pots are also useful, as they allow the sowing and subsequent planting-out operation to proceed without a setback; the roots merely grow through the peat and the whole pot can be planted in the soil. A useful tool, which has proved popular in recent years, is a 'soil blocker'. This allows a suitable, peat-based seed compost to be compressed into a block, providing a combined growth medium and pot for the germinating seed.

Most seeds will germinate fairly readily if they are given the right conditions of temperature, moisture and a suitable compost. Individual seeds will, of course, have their individual requirements. Fine seeds should be sown thinly and barely covered with compost, while larger ones are generally covered to a depth coinciding with their own width. Some seeds do not like being covered at all, so it is important to abide by the recommendations on the back of individual seed packets.

Nevertheless, everyone knows the experience of sowing seeds, doing all the right things and finding that nothing germinates. The main reason for this is insufficient warmth beneath the compost. It is possible to spend a large amount of money providing heat to warm the air in a greenhouse or conservatory and then finding that it is dissipated, without having the required effect. Germinating seeds and young seedlings need warmth from the bottom for the most effective growth. The best and most economic way of providing this is to use soil heating cables which are readily available at garden suppliers. When we first set up a small soil heating unit in our conservatory, it was a revelation to find how previously difficult seeds could be made to germinate. I had always found it virtually impossible to grow primulas from seed but, with the provision of bottom heat it became straightforward.

Soil heating cables do, of course, require a supply of electricity and, again, it is here that a conservatory

A home-made propagating unit with soil heating cables

scores over a separate greenhouse. An electricity supply is much easier to arrange in an extension of the house than it is in a greenhouse, which may be a considerable distance away. The installation of electrical power and appliances should be carried out by a competent electrician, and plugs should be on a wall, away from surfaces where water is likely to splash. The Central Electricity Generating Board provides a useful booklet entitled *Electricity in the Garden* which is recommended for all those considering the provision of electricity in a greenhouse or conservatory, or indeed, in any part of the garden.

To set up a soil heating unit, the appropriate length of purpose-made cable is bought, together with a thermostat and earthed plug. A 25 mm (1 in.) layer of sand is placed on a waterproof surface on greenhouse staging. The type of staging that we use has aluminium trays incorporated in the top, which are ideal for holding the sand. Wooden boards, 15 cm (6 in.) wide were used to make higher walls so that the unit was contained in a box-like structure. The cable needs to be laid on the sand in such a way as to form a zig-zag, without crossing over itself or touching. The closer together the zig-zags, the greater the concentration of warmth. A 25 mm (1 in.) layer of sand is placed on top of the cable to keep it in position. The sand is thoroughly moistened with water and the unit switched on ready for use. The sand must be kept moist for the even distribution of heat. Seed trays or

Soil heating system in the conservatory

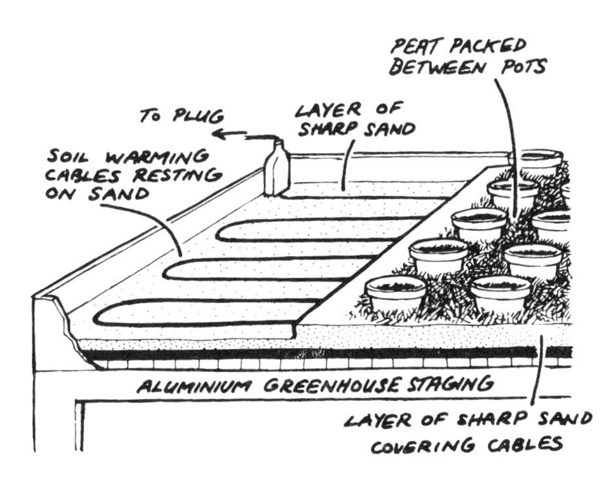

plant pots can now be placed on top of the warmed sand and sowing operations carried out in the normal way. If round pots are used, it is a good idea to place peat in between them as an insulation to stop heat escaping from the gaps. Polythene placed across the top of the boards makes the unit into a self-contained propagating unit and keeps the humidity at an acceptable level. It is also possible to purchase clear plastic propagating tops which perform the same function.

The provision of such a unit transformed our ability to produce early seedlings, but we soon ran into another snag. This was the limited amount of light. In the winter months, the number of daylight hours is considerably less than is available in the summer months. Even at midday, the sky may be so grey and overcast that little light is available. Although we had two neon strip lights in the conservatory, they were obviously insufficient to cater for the needs of some of the plants. Tomato seedlings, for example, were growing vigorously in the warmth provided for them but they were also pale green and lanky, instead of being short and dark green.

We wondered how other gardeners coped with this problem and soon found that they did not. The general reaction was that really early sowings were best left to the commercial growers. Several visits to commercial nurseries established that the growers used artificial light as well as heat, and had, in fact, been doing so for years. There was no equipment manufactured for the amateur grower which explained why the subject never appeared in gardening magazines.

It was an extraordinary coincidence that just at this time I came across a man who was just starting a small business, selling horticultural light units for the small grower. He was interested in what we were doing and kindly lent us one of his units to try out. Again, our growing activities underwent a transformation and this

Artificial light unit for the production of out-of-season food crops in the Broad Leys conservatory (Sunlight Systems Ltd)

time, we were not only able to produce the required short, sturdy and dark green tomato plants, but were also able to grow vegetables completely out of season. Salad crops were available all the year round, and we now regularly pick lettuce, tomatoes, radishes, parsley, carrots, dwarf beans in January and February when normally only leeks, brussel sprouts and parsnips are available outside.

The effect of balanced spectrum light is interesting because as well as enhancing the general growth of plants it also breaks the dormancy cycle of certain seeds which would not normally germinate out of season, even when provided with warmth and moisture. The lamp is used in conjunction with a timing unit which can be preset to switch on and off automatically. It is used to add to the existing daylight hours so that the plants 'think' it is summer. However, one interesting side effect was that we lost our African Violet plants. It was only after a conversation with a specialist grower of these houseplants that we discovered why. Apparently, African Violets need a certain amount of darkness, particularly during their non-flowering and dormant periods, and if this is not provided, the surfeit of light disrupts their cycle to such an extent that they die off.

4 Chickens

We were thrust into our first venture with chickens, with the discovery, the day after we moved in, of an old Arbor Acre hen sitting on a clutch of eggs in a box on an outhouse shelf. She was black with a flame-red comb and a temper to match. Any attempt to approach her was greeted by a raucous shriek and an outraged ruffling of feathers. We compromised, placed food and water just inside the shed and left her to it.

A few days later, six chicks hatched from eight eggs. We braved the mother's fury and brought the family down to earth to the protection of a small coop with attached run. Only two of the chicks turned out to be females; the rest were cockerels destined for the freezer. Matthew and Helen promptly christened the females

Mary, the little all-black broody hen, with yet another family

Goldie and Mary. Goldie was a peculiar bird with a golden crest of tufted feathers on her head and black feathers streaked with gold. When she grew up, she laid an insignificant number of eggs, was quite useless for anything else and was adored by the children. There could be no question of disposing of her, a situation which taught us a valuable lesson – never to give a name or make a pet of any stock likely to be slaughtered. Goldie continued for a number of years, eating to no great production until she fell foul to a predatory fox.

Mary continues still, a completely black little hen with superb mothering qualities and guaranteed to go broody several times a year. Broodiness is a quality which has been increasingly bred out of modern egg-laying breeds of hens because it reduces the overall egg yield. It is generally the older breeds which are more reliable in this respect, particularly those with a certain amount of bantam characteristics in their make-up. They suddenly decide that it is time to sit on and incubate a clutch of eggs and resist all efforts to move them. Broodiness in a hen may be inconvenient for the large producer of eggs but for the smallholder it is useful to have a foster mother ready to incubate the eggs of other breeds such as ducks, which are not always good mothers.

Bantams are small, hardy and cheap to feed. They make good pets for children

We decided early on that the distinction between pet and livestock should be clearly recognized. A second decision was that if the children were to have pet hens, it would be better for them to have bantams. These are miniature versions of larger birds and their great advantage is that they eat only about a third of the amount that larger fowl do. They are also hardy, make excellent pets and have the ability to go broody on a regular basis. We still have a small flock of bantams which is allowed the run of the site except for the vegetable garden and orchard, but they are kept quite separate from the utility birds. Only one bantam cock is kept, however, with surplus ones as well as surplus pullets being sold in the summer.

Poultry Breeds

When it came to acquiring utility chickens for eggs, we were initially influenced by our childhood memories of the breeds which had been popular with our parents. So we bought some Rhode Island Reds, Light Sussex and White Leghorns. They were big, beautiful birds – but how they ate! They were housed in traditional wooden houses equipped with perches and nestboxes and sited in the orchard so that there was grass for free-ranging. We fed them on grain and free-range pellets, a proprietary ration which provides necessary nutrients and which is formulated to balance the grass from their free-ranging activities. During the spring and summer they laid reasonably well and the big, brown eggs of the Rhode Island Reds were particularly attractive. When autumn came, the number of eggs began to dwindle until they stopped completely. We were left with a small flock of hens which did not conveniently hibernate while they were unproductive, but continued to eat vast quantities of expensive feedstuffs. To slaughter the flock was unthinkable, not only because they were such big, healthy birds, but because we would never recover the initial cost of purchasing them. We had kept a record of egg yields and feed costs but had been seduced by the production of the warm weather months into thinking that the yields were reasonable. A closer examination revealed that, averaged out over the whole year, our eggs were costing us nearly twice as much as if we had bought them in a supermarket. They were certainly fresher and better for being free-range and home-produced but the discrepancy in costs was unacceptable.

A crash programme of research into poultry history and development was begun. Our researches revealed that the traditional breeds such as Rhode Island Red, Light Sussex, Wyandottes and Leghorns had indeed been the important utility birds of our parents' generation, but time had moved on while our childhood memories had not. In the 1950s, when farming became intensive and highly specialized, the battery system of egg production was adopted. Small, commercial poultry-keepers were forced out of business, for the number of eggs they could produce with their fields of grazing hens could not compete with the production levels of environmentally controlled battery cage units. The commercial hatcheries concentrated on selective breeding programmes to produce high-yielding birds and, undoubtedly, bought up the best of the available breeding stock of traditional breeds at that time. Those that were left were the ones that no one else wanted. The small poultry-keepers who were primarily interested in 'show' birds certainly kept some of the older breeds going at a time when they might otherwise have gone out of existence, but their main interest lay in appearance not production. Because of these factors, many of the older breeds are now poor imitations of their forbears. The modern Rhode Island Red, for example, is much smaller and less prolific than that kept by our parents.

The best egg producers are now the highly bred hybrid strains such as Warren Studler and Ross Rangers which show good production levels and, at the same time, eat far less than the older, heavier birds, provided they receive balanced rations with sufficient protein levels. These hybrids are much smaller and lighter than the traditional breeds, but are of little use as table birds. While selective breeding for egg production was taking

Warren hybrid layers free-ranging in the orchard

place, there was a parallel emphasis on breeding for the table. The modern broiler bird such as the Cobb is heavy, grows rapidly, but lays comparatively few eggs. This diversification into 'layer' and 'broiler' is such that the traditional concept of a 'dual-purpose' bird is out of the question with modern hybrids.

There is sometimes confusion about the difference between brown eggs and white ones; some consumers favour brown eggs because they believe them to be free-range ones, while white ones come from batteries. This is quite untrue. Most eggs sold in supermarkets and by other retailers are from battery units, regardless of whether they are brown or white. Similarly, it is possible for completely free-ranging hens to produce white eggs. It is the genetic characteristics of the individual birds which determine how much pigment there is in the egg shells.

The Rhode Island Red breed produces brown eggs while the Leghorns have white eggs. As most of the modern hybrid strains have been developed from either of these breeds it is important to establish what sort of eggs you expect from the birds you purchase. If your choice is brown eggs, then hybrids based on the Rhode Island Red will be necessary. Hybrids developed from the Leghorn breeds will be the choice for white eggs.

The disadvantage of hybrids from the small poultry-keeper's point of view is the difficulty of breeding replacement stock. Hybrids will not necessarily breed true and it is difficult to predict what the progeny will be like. It is also extremely difficult for the smallholder to acquire a really good breeding cock, with a genetic history of good production rates. As half the potential for breeding good progeny comes from the male side, it is unwise to rely entirely on hens, no matter how good they are, if the cock is inferior. Because of this situation, the usual practice is to buy in replacement stock, either as day-olds or as pullets coming up to the point of lay, from commercial hatcheries and rearers. There is much to be said in favour of this: the birds will come from breeding stock which is guaranteed to have been blood-tested to ensure freedom from blood-transmitted diseases such as pullorum, and the chicks or chickens will also have been injected as a precaution against Marek's disease. It must be said that it is generally the small poultry-keepers who are the ones guilty of neglecting to take preventative measures to protect stock. It is not difficult or expensive to get a veterinary surgeon to carry out a blood test on poultry which is being used for breeding, and it certainly ensures that poor, weakly progeny affected by hereditary diseases are not produced.

There are many small poultry-keepers who prefer to keep pure breeds rather than hybrids because they have a genuine interest in helping to preserve the traditional breeds, and so are often prepared to put up with lower production rates as well as to lose money on feeding costs.

When we had finished our poultry research programme, we had a much more realistic picture of what small-scale poultry-keeping in the modern world involved. We decided that we would keep modern hybrid strains and would house and manage them in a way that was an intelligent and humane compromise between the traditional free-range and modern intensive methods.

Housing and Management

Traditionally, the practice was to place a hen house in a field and let the chickens free-range on the pasture, with their grass diet being supplemented by a certain proportion of grain. This is fine during the spring and summer months when the grass is in an active state of growth, but from September onwards the grass stops growing and there is little nutritional value in it. At this point, the hens were usually transferred to arable fields which had been harvested so that they could pick up any grain left over and also 'clean up' the soil of weed seedlings and insect pests. Once this was done, the birds still had winter to face and usually this meant a miserable and bedraggled existence, with a high mortality rate from cold, damp and predatory foxes. There were few eggs produced throughout the winter months, which explains why there was such an emphasis on preserving eggs in the autumn in days gone by. The usual method was to use 'waterglass', or sodium silicate dissolved in water. The idea of the fresh, free-range eggs of the 'good old days' is indeed a myth. Most people over 40 will remember the high incidence of bad eggs, even amongst shop-bought ones – a situation which now no longer exists.

We had no wish to follow barbaric practices such as those we had seen in the batteries. Whatever method of management we followed, we were determined that it would respect the natural instincts of the birds – to peck, to scratch, to perch and to take dust baths. So, the problem was to arrive at a system that was humane but which still enabled us to have eggs throughout the year.

We decided that the stable would make an excellent poultry house. It had electric light, in the form of a neon strip light, and a two-part stable door. All that was necessary was to put up some perches, about 75 cm (2½ ft) off the floor, to suspend a feeder and drinker, provide nestboxes and suitable litter for the floor, and to make a 'pop-hole'. The latter is a hole in the south-facing wall fitted with a small door which can be raised or lowered to open and close as desired. It is very important to confine the chickens at night otherwise they are at risk from predators such as foxes. Where the pop-hole opens on the other side of the wall, we constructed a small run. This opens out at one end with access into the orchard (see diagram).

The thinking behind this system was firstly that the house would provide permanent sleeping accommodation. During spring, summer and early autumn, the chickens are allowed access into the orchard, via the pop-hole and run, so that they have the benefit of grazing and free-ranging. Each evening they return to

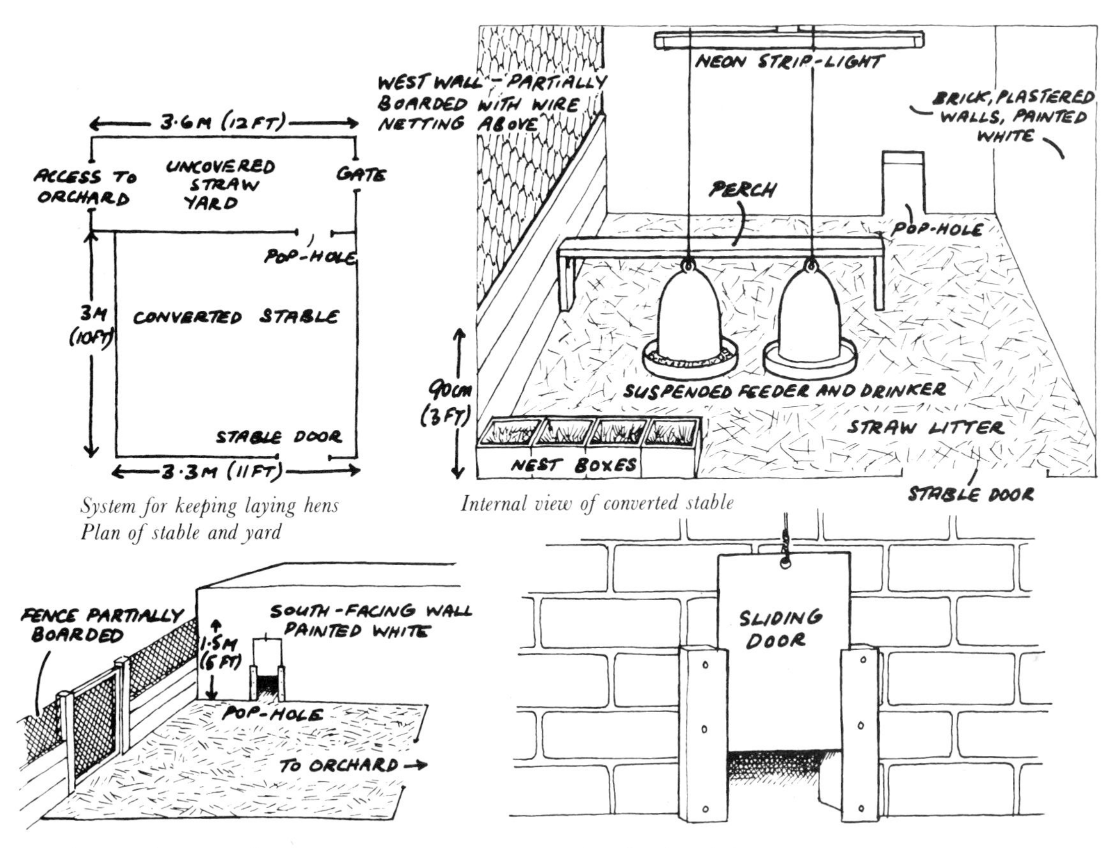

System for keeping laying hens
Plan of stable and yard

Internal view of converted stable

View of uncovered straw yard

Details of pop-hole

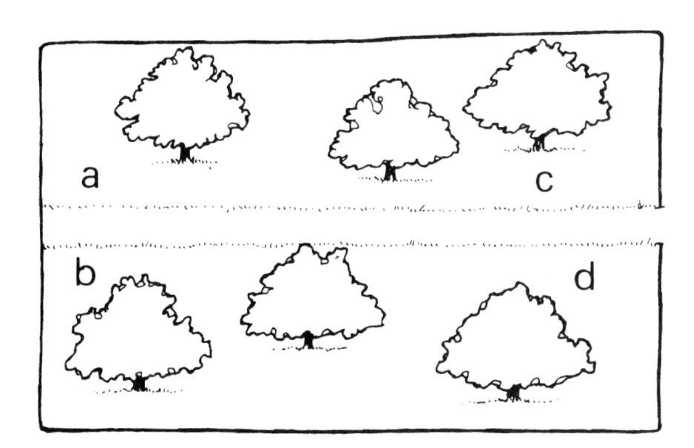

Rotation of grass in the orchard at Broad Leys
(a) Position of fence for following year
(b) Temporary post and poultry netting fence
(c) Grazing area 1 – in use. There is access to the grazing area from the yard and the house (closed in winter)
(d) Grazing area 2 – resting. While not in use, the ground is raked to break up droppings and limed to get rid of parasites

the house to be locked up for the night. If the weather is particularly bad, such as a period of continuous rain, they are not allowed out of the house at all – otherwise they quickly become soaked and bedraggled. It must also be admitted that chickens are not the brightest of beings and will often not have the sense to seek shelter when the rain comes. Any drop in temperature means that more of their energies are required to keep the body metabolism at normal level, and this is often accompanied by a drop in egg production.

Once the grass stops growing in late autumn, there is no point in letting the chickens go out into the orchard. There is little food value in the grass and it needs to have a period of rest in order to recover for the next season. The hens are confined to the house and run during this period. The run enables them to have fresh air and exercise in a sheltered area; it is against a south-facing wall which is painted white in order to reflect as much of the winter sunshine as possible. We put straw down in the run to prevent boggy conditions.

In the house itself, natural light is balanced by the appropriate amount of artificial light. The hens need a minimum of 15 hours of light a day in order to continue laying, so it is a matter of working out how many extra

Interior of a layers' house

hours of electric light are required to make up the total. Light can be provided before dawn, after dusk, or a mixture of both. When we first started with this system, I wrote in the magazine that I was the one who had to get up at 5 a.m. in order to switch on the light. A few days later a small parcel arrived. It was a reconditioned timer which a reader had sent in, with instructions for us to insert it into the circuit and programme it so that the light came on and went off automatically. With the timer was a suggestion that we give him a free subscription to the magazine in return. It seemed to be in the best traditions of barter and we were happy to agree. Now, several years later, we are still using the same timer and it works perfectly. There are several poultry equipment suppliers who sell these timers and the most sophisticated of them also have a dimming device. This is so that there is a period of dimming before the light is extinguished, warning the hens to go to their perches; otherwise they would be plunged into sudden darkness and could not find their way.

When it came to the provision of nestboxes in the house, the chickens had their own ideas on the subject. I carefully constructed a bank of nests from a series of old boxes and placed them along the base of one wall. The hens ignored them and laid their eggs in a great heap in the opposite corner. In fact, there was great competition for this particular spot and it was not unusual to see about four birds all sitting virtually on top of each other. There was no alternative but to adopt the 'mountain to Mohammed' philosophy and move the nestboxes to that particular area. One nestbox is required for every three hens and clean straw or wood shavings should be placed in them to make them attractive to the birds. This also needs frequent changing to ensure that the eggs are laid in clean conditions.

Wood shavings or straw are also needed as litter on the floor of the house in order to absorb the droppings. The litter does not need to be replaced immediately it becomes soiled, but can have a fresh layer added on top as required. Clearing out once every three months is adequate and our practice is to close the pop-hole to exclude the chickens from the house while this is done. The litter is forked and spaded into a cart and the floor then brushed with a stiff broom. Lime is sprinkled as a deterrent to pests and a fresh layer of clean litter applied. We also take the opportunity of dusting the perches and nestboxes with a proprietary insecticide to deal with lice and mites, external parasites which can be a nuisance on poultry. In fact, the chickens themselves are given a dusting before being put back into the house. Particular attention needs to be paid to the areas under the wings, around the vent and on the neck.

The litter which is removed from the poultry house is added to the compost heaps where it breaks down into a valuable compost, high in potash, for the vegetable garden. The grass in the orchard is raked to disperse any droppings, then given a liberal dressing of lime to get rid of any residual parasites such as tracheal or gizzard worms.

The system that we have evolved for our chickens works well. It ensures that we have eggs throughout the year, the chickens are protected from the bad weather and their natural instincts are respected. They are fed twice a day, in their house, and if we need to be away for a few days there is no problem. It is simply a matter of filling up their feed hopper and drinker and then confining them to the house. The feeder and drinker are of a sufficiently large capacity to last for several days without needing replenishment, and the fact that they are always kept inside the house means that wild birds do not feed at our expense.

Feeding

Feeding is an area where it is possible to lose money unless costs are carefully noted. A hen will eat between 80–160 g (3–6 oz) of food a day, depending on her size, age and level of egg production. She will need a balanced ration of proteins, fats, carbohydrates, vitamins and minerals, as well as unrestricted access to fresh, clean water. Proprietary feed is available, which provides all the necessary nutrients except water. It is obtainable in the form of pellets or dry mash which can be mixed with water to a crumb consistency. This type of ration was developed as an 'all-in', convenience feed for the large, commercial units and is expensive. The smaller poultry-keeper will find it more economic to feed a proportion of proprietary feeds and make up the difference with grain, grass and some home-produced feed.

Our feeding pattern is to give pellets to the layers in a feed hopper every morning, and wheat in the evenings. During the day, the layers have access to grass for grazing in the orchard as well as having a proportion of vegetable peelings, and kitchen scraps placed in their run. On really wet days, when they are confined to their house and cannot graze, they are given the outer leaves of vegetables such as brassicas to peck. In winter, when there is no grass available, it is particularly important to provide greens for them, not only as a source of minerals, but also to alleviate boredom which may otherwise trigger off abnormal behaviour patterns such as feather and vent pecking of other birds or egg-eating, and it is important to ensure that sufficient levels of calcium are available in the diet. A traditional practice is to bake egg shells, crush them and feed them back to the birds. This works well, but it is important to make sure that they have been heated sufficiently to sterilize them, and then crushed finely enough so as to be unrecognizable as a shell to the bird. Proprietary grit, as a source of calcium, is available from poultry suppliers. The other essential in the diet of chickens which eat grain is crushed oystershell or small pieces of gravel. These are taken into the gizzard, an organ with strong muscular walls which is responsible for grinding up hard grains and seeds. Unless crushed oystershell or gravel is available, the grain cannot be digested properly. Free-ranging birds will usually find their own small pieces of gravel, but it is a good idea to play safe and place a small dish of crushed oystershell in their house or run. They will help themselves as necessary and it will only need topping up once every few months.

Raising Chicks

I have already said that it is advisable to have breeding stock tested to ensure that they are free of blood transmitted diseases which might adversely affect the progeny. If good layers are required, then it is necessary to have a cock and a hen which both come from lines with a good record of laying. Where hybrid strains are concerned, it is difficult for the small poultry-keeper to obtain a good breeding cock and it may, therefore, be necessary to buy in replacement stock, either as day-olds or point of lay pullets. This is a question for the individual smallholder to decide.

Incubation of eggs can either be carried out with an incubator or using the services of a broody hen. Broodies provide the best and most natural method, as well as the necessary aftercare, but unfortunately, tend to go broody at inconvenient times. Just when a clutch of fertile eggs is ready for incubation, is always the time when there is a dearth of broodies. Small incubators, catering for about 50 eggs at a time, are available. Our practice is whenever possible to use broody hens, but when these are not available, to fall back on the electric incubator. Incubators can, of course, be used to incubate the eggs of all fowl, so if a range of different poultry is kept, then it is worth the investment. We have used ours to hatch the eggs of geese, ducks and quail, as well as chickens. Although the incubation period varies, the basic necessities are the same.

Eggs which are intended for incubation must obviously be fertile, and should be incubated as soon as possible after laying. Having said this, it is not always possible to have all the eggs ready at one time; they are laid at different periods. While awaiting incubation, the eggs may be stored in a cool pantry, ideally at a temperature of 12–15 °C (55–60 °F) and should be turned every day. Once a clutch is available and before any of the eggs are 14 days old, they should be placed in the incubator, with a cross marked on each egg. This is to ensure that regular turning of the eggs is carried out, ideally three times a day. The incubator should be set in such a way as to provide a temperature of 39·4 °C (103 °F) when a thermometer is held 5 cm (2 in.) above the centre of the eggs, and the water tray should be kept topped up to ensure adequate humidity. It should be allowed to run for 24 hours before the eggs are introduced, so that the temperature can be checked and, if necessary, adjusted.

A broody hen on a clutch of eggs. When they hatch, the chicks will be put with the mother hen in a confined run

Chicken's eggs take 21 days to incubate and, if all goes well, the chicks will begin to 'pip' at any time from the twentieth day onwards. 'Pipping' is the cracking of the shell by a specially strengthened section of the beak, rapidly followed by the emergence of the chick. The incubator should not be opened until all the chicks have emerged and their feathers have dried off. They will not need food for the first two days because remnants of the yolk are still available in the abdomen – this is why it is possible for hatcheries to despatch cartons of day-old chicks with such success.

There are bound to be some eggs which will not hatch, either because they were not fertile in the first place, or because of accident resulting in death in the

An inexpensive incubator suitable for hatching a small number of eggs (Mardle Incubators)

shell. It is always best to remove any obviously infertile eggs as soon as possible in case they go bad and infect the remaining eggs. It is possible to detect which eggs are fertile, after a week in the incubator, by a process called 'candling'. This is the examination of an egg in a darkened room, allowing bright light to pass through it. A fertile egg will be seen to have a developing embryo, which will appear as a reddish blob with star-like fingers projecting from it. Examination should be carried out as quickly and as gently as possible, before returning the fertile eggs to the incubator and discarding the others.

Once the chicks have all hatched, or in the case of bought-in day-olds, been received, they should be placed in a protected environment for brooding, or raising until they are weather-hardy. A broody hen should have a protected run, for herself and her chicks. This will need to be equipped with a feeder and drinker.

Bought-in chicks will need an artificial brooder, in other words, suitably protected conditions where cold, dampness and rats are excluded. On concrete, a thick layer of wood shavings provides a warm, insulated floor. An infra-red lamp, readily available from suppliers, will keep the chicks warm, while temporary walling of some kind will keep them confined in one small area where the effect of the lamp is felt. This walling could be of any readily available material and many people find that corrugated cardboard is as good as anything. If there is any possibility of rats getting into the outhouse, however, the brooding unit will need to be made of much more substantial materials. Purpose-made brooders are also available from commercial suppliers.

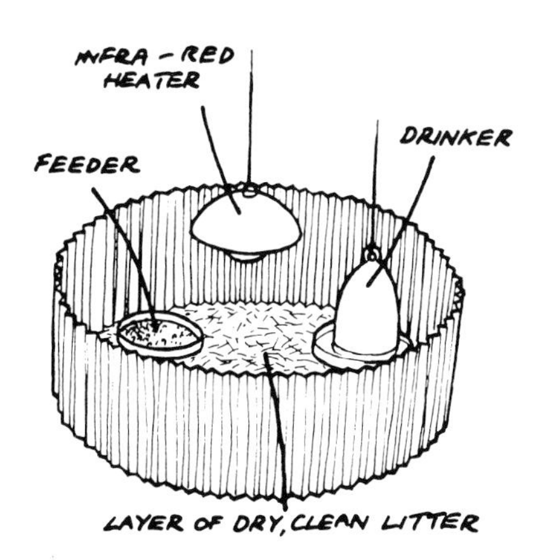

Small brooding area in protected building. Temporary walls are used for confining the chicks near the lamp. The floor should be dry and rat-proof

The position of the lamp is important and can be gauged by watching the behaviour of the chicks. If they huddle together in a mass in the centre, they are too cold and the lamp should be lowered. If, however, they are scattered to the edges of the confined area, they are too hot and the lamp should be raised slightly. The only other provisions necessary at this stage are food and drink. Proprietary chick crumbs will give them the best start in life and should be made available in a suitable container, so that they can help themselves at any time. Fresh, clean water should be available, in an enclosed drinker so that they cannot climb into it and get wet and chilled, or possibly even drown.

As the chicks get bigger they can be given more room, first by moving the temporary walls further apart and, eventually, dispensing with them entirely. The lamp will only be needed until they obviously no longer need it. This will depend on weather conditions. If it is particularly warm, the chicks can be allowed out on grass after a week or two, but only in a confined run which gives them protection against predators. They can be given chick crumbs until they are about seven weeks old, when a different food such as a proprietary grower's ration and wheat can be introduced gradually. Crushed oystershell should also be given for the proper digestion of grain. At this stage, the diet will depend on the type of chicks. Those that are being raised for the table will have a different pattern of feeding (see page 57) from those scheduled for laying which will be fed less intensively at this stage. If they are grazing, a grower's ration can be given in the mornings, with wheat, or a mixture of different grains given in the evenings. Once they get to 'point-of-lay' at about 20 weeks, they should be introduced to a proprietary layer's ration in the mornings, to replace the grower's meal.

Culling Stock

Any chicks which turn out to be cockerels will need to be culled at 10–14 weeks. So will females which show no signs of coming into lay, or older hens whose egg production is declining. The signs that a hen is in lay are: a full red comb, a rounded, moist vent, as distinct from a conical-shaped, dry one, and an appreciable gap between the pelvic bones. This can be established by seeing how many finger widths there are between the bones, as shown in the photograph. One finger width indicates that she is definitely not laying, two means that she could be and it would be wise to do a further check in a few days' time. Three finger widths indicate that she is definitely in lay.

Any stock which is due to be culled should be kept confined in a protected place and not given food for 12 hours, although water should be freely available. Neck dislocation is the traditional and easiest method of killing, but should not be attempted without first being demonstrated by someone experienced. The technique is to twist and pull the neck simultaneously so that it is broken. A certain amount of wing flapping will result, for there is always a nervous reaction for several minutes, even after death has taken place.

Plucking should take place immediately after slaughtering because the feathers come out more easily while

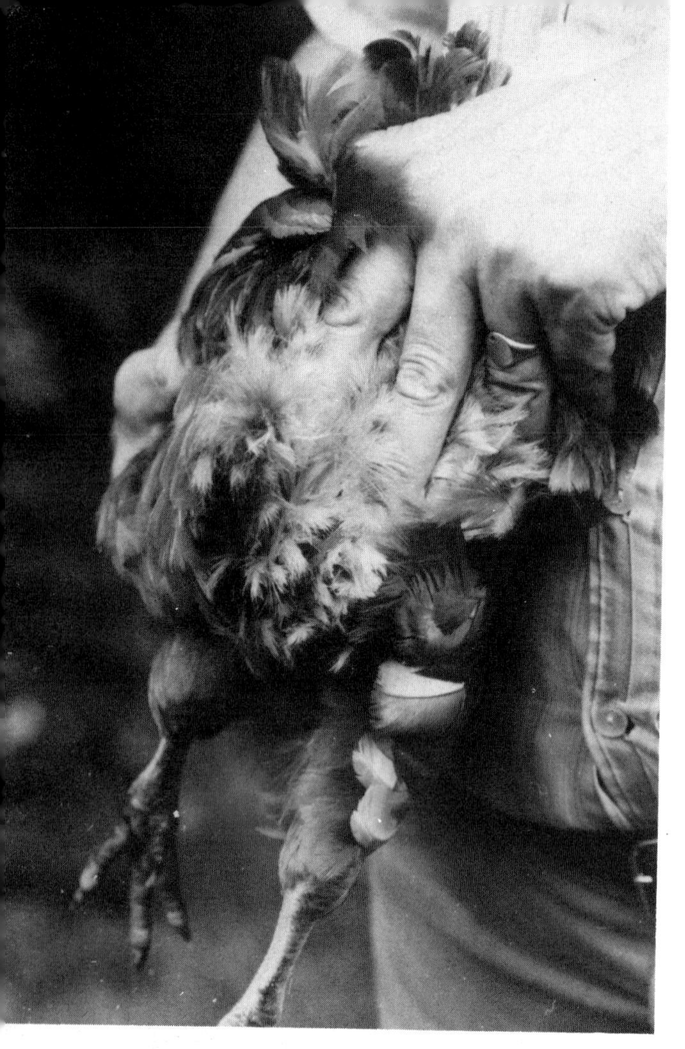

Checking the width between a hen's pelvic bones to find out whether she is laying

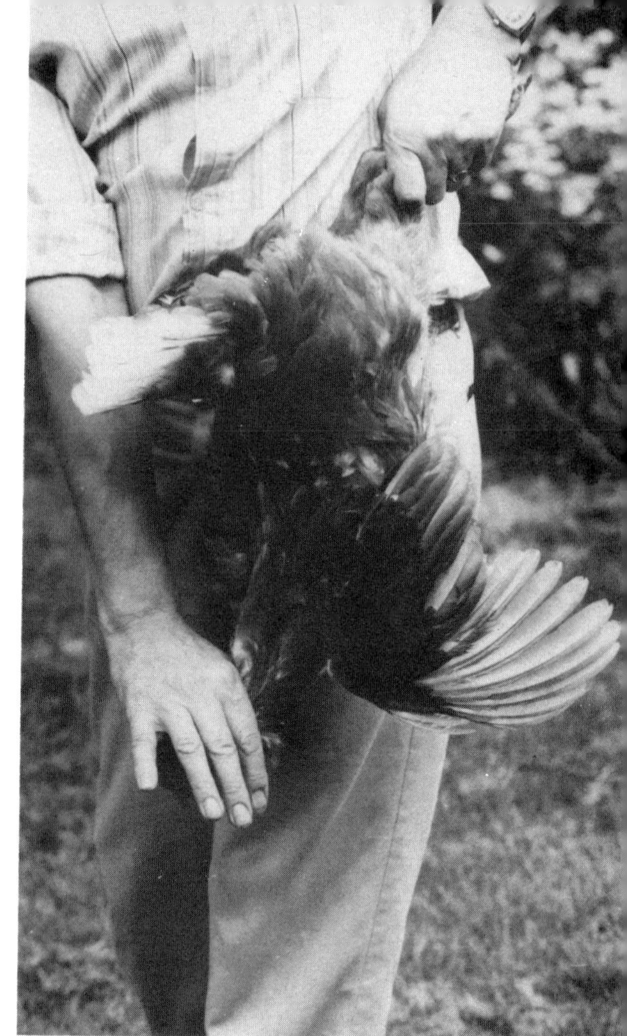

Killing a chicken

the carcass is still warm. It does not matter in which order the plucking takes place; some people prefer to pull the body feathers out first, while others prefer to start with the wings. The principle is to pull the feathers sharply, 'against the grain', but without tearing the flesh.

Once plucking is complete, the birds can be suspended, head-downwards, in a cool room. Hanging for a day or two before gutting does improve the flavour of the meat, and the period is largely dependent on the outside temperature and on personal taste.

Gutting is not difficult and is a matter of removing the innards, without breaking the bowels and making a mess. Cut off the head with a sharp knife, then sever the neck at the point near the bird's 'shoulders'. It will snap off and can be removed by holding it with a clean cloth and pulling it firmly. Put your hand inside and loosen all the connective tissue and tendons. Now turn to the other end and with a sharp knife cut around the vent until it can be gently pulled out. Use the fingers to break the surrounding connective tissue, and the vent with attached bowels can then be drawn out in one piece. Take care not to rupture the intestines. Make the incision big enough to allow the whole hand to go inside and grasp the remaining giblets, pulling them out. Any that are inaccessible from the vent end can be removed from the neck end. The bird is now ready for cooking or freezing. Chickens up to the age of one are suitable for roasting, but older birds should be boiled, otherwise they will be too tough to chew. Boiling makes them extremely tender.

Raising Table Chickens

Surplus cockerels and laying hens can be culled for the table, but if really big table fowl are wanted, it is best to raise one of the specially developed table breeds such as the Cobb. These are big, heavy birds which have been selectively bred for rapid growth.

There is little point in acquiring these birds and then

letting them free-range for they will run off all their weight and you will end up with expensively fed, inferior birds. They are best kept confined until they reach killing weight. This does not mean that they need to be kept in cramped conditions, in semi-darkness to keep them docile as the commercial broiler producers do. The ideal is a simple house with perhaps one wall covered with wire netting so that they have plenty of fresh air and natural light. As our turkey house is only used from late summer until Christmas, we use it to house Cobbs for the freezer in the earlier part of the year, from spring to mid-summer. Details of its construction and use are given on page 75. As far as feeding is concerned, a broiler ration can be given on an *ad lib* basis, or a more varied diet can be arranged. The drawbacks of broiler rations for the smallholder are the expense and also the fact that they contain antibiotics, not only to counteract infection such as coccidiosis, but also to accelerate growth. Where the birds are kept in a more humane, less intensive way, there is no need for these additives, and medical and veterinary authorities have frequently condemned such indiscriminate use of antibiotics. A suitable alternative fattening ration would be a mixture of wheat, oats and barley.

There is no set time for deciding when to kill the birds for the table, as long as they are not allowed to go into the winter months. After this time, they will begin to lose condition and feeding costs will escalate. Apart from this, it is a matter of deciding which weight is the most appropriate for your oven. Cockerels will be heavier than females. It is a good idea to have a spring balance which can be suspended from a beam, so that a periodic check can be kept. It is simply a matter of tying a strip of cloth around the legs of one of the birds and hooking it onto the hook of the scales. Watch out for the fluttering wings and, as soon as the weight is noted, release the bird. Anything in the range 1·4–3·7 kg (3–7 lb) live-weight makes a good table bird.

Health

I have already stressed the importance of acquiring blood-tested and healthy stock from a reputable supplier. As far as the smallholder is concerned, the most important aspect of poultry health is avoiding trouble. Feeders and drinkers should be kept clean and houses regularly cleared out and, if necessary, fumigated.

Flooring litter, whether wood shavings or straw, should not be allowed to get damp, otherwise there is a danger of coccidiosis. Perches and nestboxes should be periodically dusted with a proprietary lice and mite powder, and the birds themselves should be treated in this way once every six months. Table birds will not

Table cockerels ready for slaughter

normally require this for, with a shorter life span, they are unlikely to be badly affected. A careful watch should be kept for the condition of scaly leg, indicated by white encrustations pushing up the scales of the legs. This is caused by the burrowing action and secretions of a mite. The treatment is to scrub the legs with an old toothbrush and warm soapy water, then, after drying, to paint them with benzyl benzoate, obtainable as a proprietary compound from a veterinary surgeon.

Free-ranging birds are susceptible to parasitic worms of the caecum, gizzard and trachea. The best control is to provide fresh, clean grass which has not been grazed by other poultry the previous season. A system of rotation of grass is essential, and new stock should not be allowed to mix with old stock. The method that we use at Broad Leys is to divide the orchard into two, so that one half is grazed, while the other rests. As soon as hens are confined in their house and run for the winter, the vacated area of orchard is raked to disperse droppings and limed to kill off residual parasites. The following spring, the hens are allowed to graze the other half. The bantams and 'pet' hens are kept quite separate. They are allowed to free-range everywhere except the vegetable garden, orchard and field. They are not rotated, but as we have extensive lawns, and their numbers are small, there is no significant build-up in the worm population. The layers are kept for two years before being slaughtered and given to a local man for feeding to his ferrets. After cleaning and disinfecting the stable, a new set of point-of-lay pullets is introduced. The layers are wormed twice a year but not the table birds, for they are not free-ranging and do not have time to become infested. The easiest way to administer a poultry vermifuge is to add it to the drinking water. Keep the birds confined in a house without water for about six hours, then introduce the dosed water and they will all drink.

At the first sign of illness in a bird, it is good practice to separate it from the others in case the infection spreads. Keep it in a warm, sheltered place with access to food and water and observe it. If it is not a serious condition, the bird will probably improve. If it dies, the odds are that it would have done so anyway, regardless of any treatment you might have given. It is not worth calling a vet to attend a single bird, for that would be economically unsound, but if several birds show similar symptoms, then get immediate advice. There are two notifiable diseases which affect poultry – fowl pest and Newcastle disease.

Any dead birds should be incinerated, rather than buried, in order to minimize the spread of infection. Obviously no sickly bird should be killed and eaten.

Finally, one of the worst carriers of disease is the rat, and every effort should be made to control the numbers. Erradication is impossible, but much can be done to keep them under control. Make sure that all feedstuffs are in ratproof containers; surprisingly enough plastic dustbins with lids are safe, while metal ones are obviously suitable. The local authority will provide a free pest control service to any householder in its area, but registered smallholders will have to pay. Private or non-registered smallholders are regarded as householders and, therefore, eligible for free service.

Hens take frequent dust baths to help in getting rid of external parasites

5 Ducks and Geese

Ducks

When we first arrived at Broad Leys, we inherited a small flock of Khaki Campbell ducks. This was the breed which was kept by my parents and indeed by all those interested in egg production. First bred by a Mrs Campbell in Gloucestershire, they quickly replaced the Indian Runner as the best producers of duck eggs. They are not good table birds, however, and it is the snow-white Aylesbury which has traditionally had this role in Britain. In France the Rouen was the prominent table duck, while in China, the USA and Australia, the Pekin was most widely kept.

All ducks were originally developed from the wild Mallard and share the same basic characteristics, but there has been a wide divergence into domestic types for eggs, meat and down feathers and ornamental breeds which are kept for their appearance. The Call duck, a small white bird with a piercing call, was bred as a decoy for wild ducks. It was used to lure them onto wide stretches of water where shooting parties with retriever dogs would be waiting.

There is a wide choice of breeds available, but for the smallholder, the best choice is to concentrate on domestic types which have been selectively bred for commercial purposes. If eggs represent the main priority a good strain of Khaki Campbell is the most appropriate. For table birds, the best choice would be the Aylesbury, while a good dual-purpose bird for eggs and the table is the Welsh Harlequin. It should be emphasized, however, that merely acquiring any of these breeds is no guarantee; they must be good 'strains' which have been selectively bred for production, rather than looks. This means going to the commercial breeders, rather than to those whose main interest lies in exhibiting.

A good strain of Khaki Campbell will lay more than 300 eggs a year and is better than the best of hens, but, in order to attain production levels of this sort, they need to be properly housed and adequately fed. The ones we inherited were virtually wild and had been allowed the run of the site. They had no house and were fed only scraps from the kitchen window, with the occasional handful of wheat. They had bred quite indiscriminately and there was a large number of surplus males who were always fighting. As far as we could establish, the numbers had been controlled only by periodic visits from foxes.

There is nothing to be gained by keeping domestic ducks in this way. Ducks will lay all over the place and must therefore be housed until mid-morning so that the eggs are in an accessible place. When we first arrived we wondered why there were so many magpies around, and then soon realized that they were drawn to the site because of the duck eggs left all over the lawn, like white mushrooms in the morning. Magpies are sharp-eyed in this respect and will swoop down, crack open the eggs and eat the contents. Even the pond, where the ducks spent a large amount of their time, had its quota of eggs bobbing around. The children used to refer to it as the 'duck-apple' pond. Such eggs are quite unsafe to eat because of the danger of salmonella infection. Duck eggs have quite big pores in the shells and absorb dirt and bacteria more easily than hen's eggs, particularly if they are laid in damp, dirty conditions. It is absolutely essential to provide laying ducks with clean, dry nest-boxes in which to lay and to change the nesting material frequently.

The question of whether ducks need a pond is often raised. It is certainly a natural environment for them, but not absolutely essential, as long as they have water deep enough to immerse the head and neck and to dampen the feathers. Table ducks which are raised to killing weight in 10–12 weeks never have access to water for swimming. If a pond is available, laying ducks are certainly happier if they have the use of it, but every effort should be made to keep it clean and to remove scum and decaying vegetation. The latter may harbour a form of botulism which causes paralysis in the ducks after they have been dabbling in it. Severe cases are fatal. If there is any question of sewage outfall being present, as there used to be in many old farm ponds, the ducks should certainly be excluded because of the risk of disease transmission via the eggs to the consumer.

Housing and Management

We decided that as our pond is fairly clean and with no contamination from any source, the ducks should continue to use it, but we would provide a house for them, whether they liked it or not. There were a number of old, fairly ramshackle poultry houses on the site, and

Welsh Harlequin ducks and the pond at Broad Leys

A small pond is appreciated by ducks but it must be kept clean if they are egg producers

one of them was repaired and adapted for duck occupation. It had a slatted floor through which droppings could fall, a door, pop-hole and window, and the wooden walls were topped by a sloping wooden roof. This needed new roofing felt to make it waterproof, for, despite the fact that they are waterfowl, ducks need dry conditions in which to sleep.

There were no nestboxes, so these were provided for them. Ducks do not use nestboxes as readily as hens and, as mentioned earlier, do have a tendency to drop their eggs anywhere. At the same time they also tend to use nests, if they are available, although there are always a few culprits who lay on the floor. The boxes we made were sited along one wall of the house and were shallower than those normally used by hens. Ducks' feet are not built for jumping and perching so they dislike sharp increases in height. For the same reason, a ramp was provided for them outside the pop-hole so that they would not strain their legs in having to jump up or down. Straw was placed in the nestboxes and a drinker suspended from the roof. A large, shallow, cast-iron, and therefore extremely stable, feed trough was placed on the slatted floor. Finally, new wire netting was nailed across the window space so that foxes were excluded, but ventilation was ensured.

The reasoning behind this system was that the ducks should have night protection and a clean, accessible place for laying. Most of the droppings would fall through the slatted floor and any which did not could be brushed through with a stiff handbrush the following morning. Once every few months, the whole house could be lifted so that the accumulated droppings underneath could be swept up and put in the compost heaps. Food is given in the house in the late afternoon, as an inducement for the ducks to go in, while water is obviously necessary as an accompaniment. They are then confined in the house until mid-morning the following day, so that all the eggs are laid inside. These are collected when the ducks are let out.

Once the house was ready, we sorted out the flock and slaughtered most of the drakes, leaving the two best ones, so that there was a breeding ratio of one drake to every five ducks (see also page 64). It took quite a few days to persuade the rest to start using the house and it was only achieved by withholding food completely, except in the late afternoon, when a great show was made of filling the feeder in the house. Occasionally, one duck would almost be in, then would panic and flap its way into the pond from where it was impossible to retrieve it. At last, just when we thought that our patience was exhausted, they understood the message, 'no house, no food', and our problems were over. It is worth mentioning that whenever new poultry, of any kind, is introduced on a site, the birds should always be placed in a house with food and water, and kept confined for at least 24 hours and ideally two days, so that they learn that that is their home. If they are then released they will go back to the house in the evening because they associate food and rest with the house.

Feeding

Ducks scoop food up in their wide bills so it needs to be placed either on the ground or in a shallow feeder. Water needs to be available nearby so that they can take frequent drinks in case dry food clogs up the nostrils.

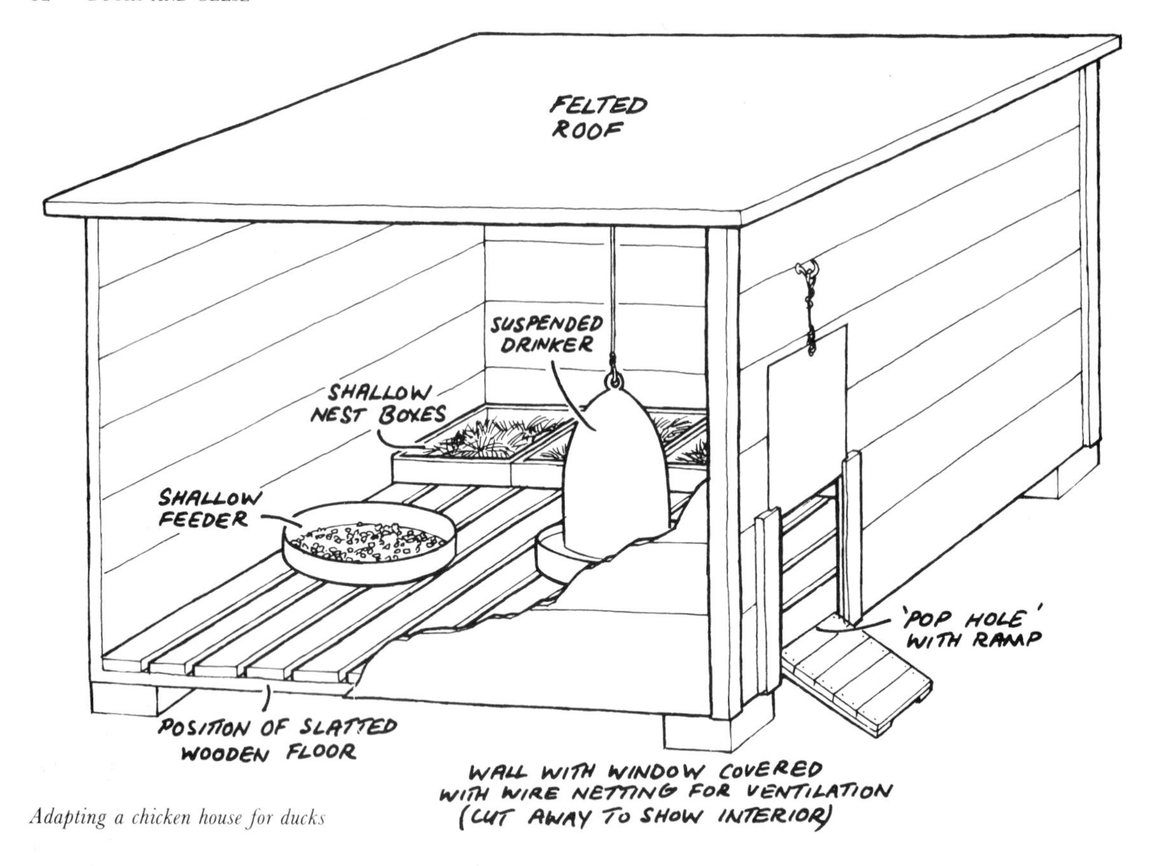

Adapting a chicken house for ducks

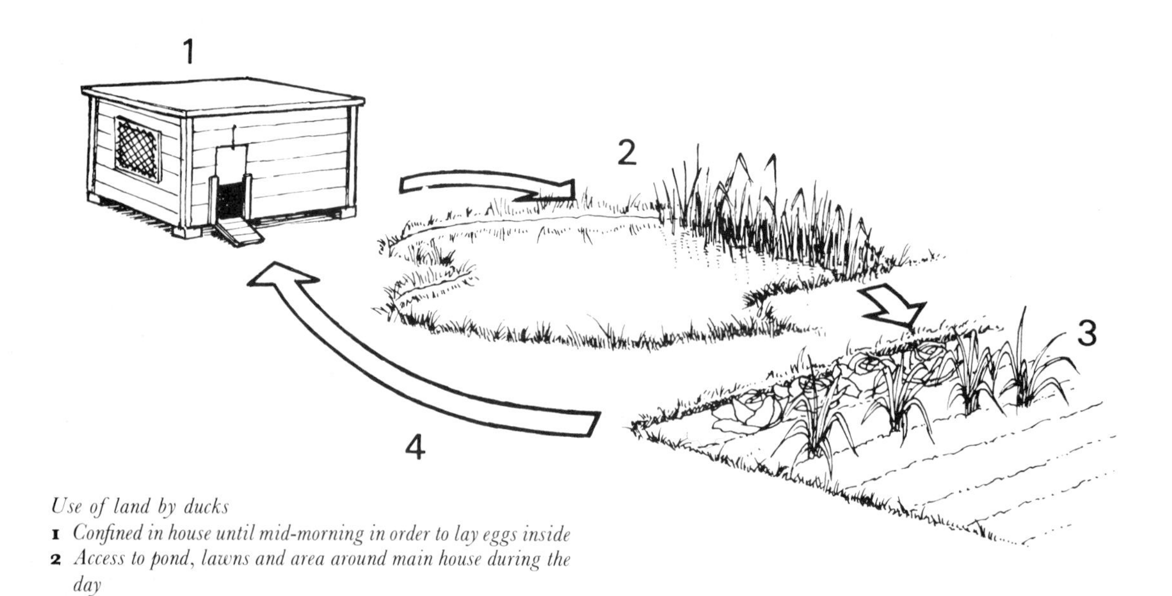

Use of land by ducks

1 *Confined in house until mid-morning in order to lay eggs inside*
2 *Access to pond, lawns and area around main house during the day*
3 *Access to the kitchen garden for slug clearance in winter only*
4 *Fed in house in late afternoon and locked up for the night*

Laying ducks can be given layers' pellets such as those produced for hens and an appropriate time to give these is when they are let out in the mornings. We find it more convenient at this time to feed them from a scoop on the ground outside; they quickly eat the pellets up, then make a dash for the pond. In the evenings, wheat is given in the feeder in the house. During the day, the ducks forage in the pond and on the lawns around the house, but they are denied access to the rest of the site.

Rearing Table Ducks

Khaki Campbells are rather skinny birds and not really worth rearing for the table. It is much better to concentrate on a big, heavy breed like the Aylesbury. The problem with having two breeds is in keeping them separate, otherwise they will interbreed. After vainly trying to achieve this, we sold our whole stock of ducks and invested in a breeding stock of dual-purpose Welsh Harlequins. These are big, placid and friendly ducks which are good layers and make excellent table birds.

We housed them in our existing duck house and as they bred, a proportion of them, not needed as layers and including surplus drakes, were transferred to a separate house for fattening. The house we use is the turkey house (see page 75). As this is vacant in the first half of the year, it is put to good use.

Table ducklings do well on a fattening ration of wheat, stale bread and barley, with drinking water available at all times. The killing time is somewhere between 10 and 12 weeks and is really dependent on the rate at which they put on weight. (Commercially, killing time can be as early as six weeks). The crucial factor is when they have their first moult and acquire their new feathers. The ideal time to slaughter is just before the new feathering starts, otherwise there will be large numbers of feather stubs in the carcass. The methods for killing and gutting are similar to those used for table chickens. As the breast or down feathers are a valuable commodity for stuffing cushions and duvets, it is worth keeping these separate from the rest of the feathers at plucking time, and storing them in a large pillow case until needed. They are best stored hanging up so that air can circulate – pegging the pillow case on a line in an outhouse is ideal.

Welsh Harlequin drake with two of his ducks. This is a good dual-purpose breed for the table and eggs

Breeding

It was mentioned earlier that for a light breed such as the Khaki Campbell, a breeding ratio of one drake to every five ducks is appropriate. For a heavier breed such as the Aylesbury or Welsh Harlequin, one drake to three ducks is more suitable. The onset of sexual activity is demonstrated by a characteristic head bobbing between the duck and drake. Mating can take place in water or on land and there are some who claim that fertility is increased when the former occurs, but there is no available evidence to substantiate this. Generally speaking, ducks do not make good mothers, although there are obviously many exceptions to this. What is certainly true is that there is less incidence of broodiness, hatching and successful rearing with ducks than with hens. Most people who keep ducks prefer to use broody hens or incubators to hatch their duck eggs.

An interesting alternative is to use Muscovy ducks as broodies. No one is quite certain whether the Muscovy is a duck or a goose, for it displays characteristics of both. What is beyond question is that it is frequently broody and makes a good mother. We acquired some of this breed purely to act as broodies and they were most successful. Although Muscovys will mate with other ducks they do not appear to produce young from such matings. The snags were that they tended to be aggressive to the other ducks and had a disconcerting tendency to fly. After a short period we sold them and acquired a small, electric incubator.

The average incubation period for domestic ducks is 28 days, and, once hatched, the ducklings should be given the same care as that described for chicks (see page 55). A broody hen or duck with ducklings will require a protected shelter and small run. There is nothing to be gained from letting them wander freely where they will be at risk from predators, and on no account should they be allowed to go on the pond until they are six weeks old, and properly feathered. A mother duck will take her brood onto water and be quite oblivious of the fact that they are being decimated by cold. In nature, the mortality of ducklings due to rats and cold is extremely high.

Ducklings do well on proprietary chick crumbs which

Muscovy ducks make excellent mothers but are inclined to be aggressive to other birds

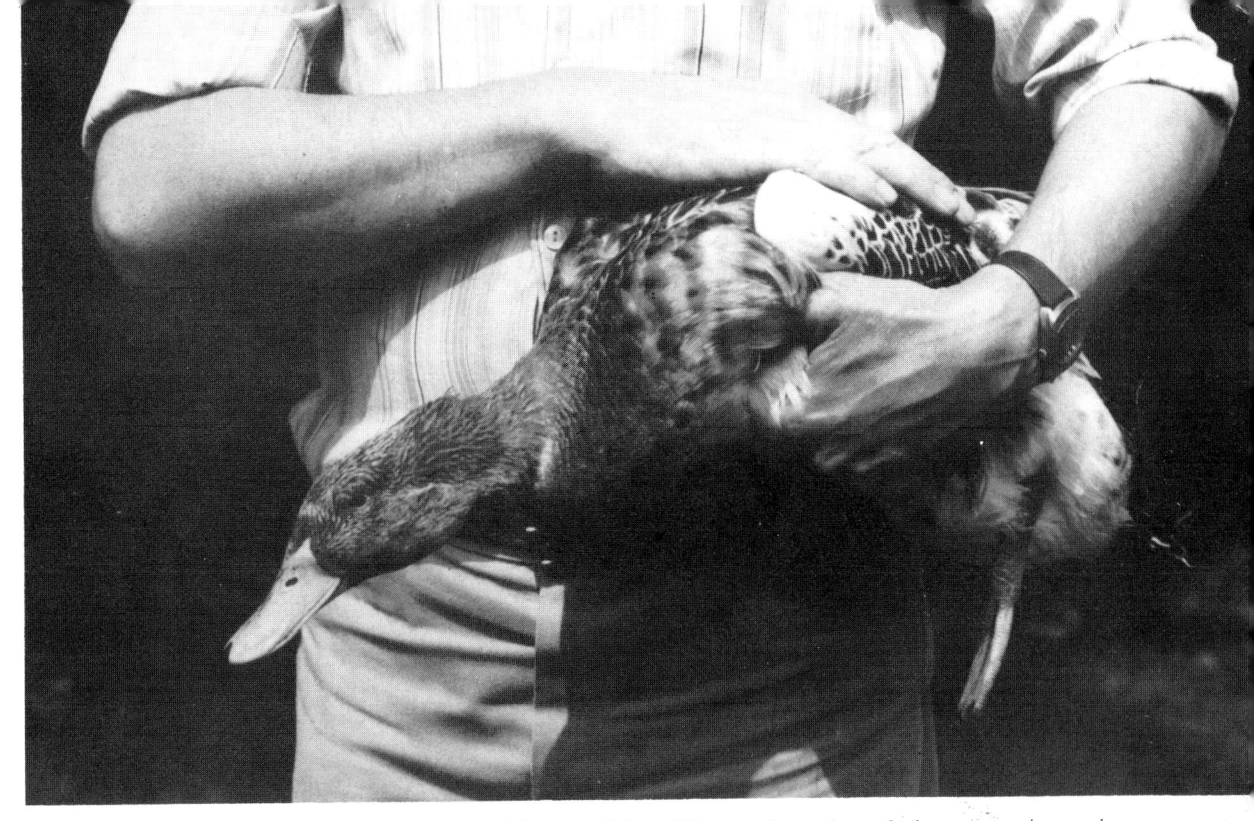

Carrying a duck. It should be supported from below and the wings kept confined

Below *The tips of the primary feathers on one wing can be clipped in order to confine ducks*

are high in protein and give them a good start in life. From about six weeks onwards, they can be given the appropriate adult diet, depending on whether they are laying stock or scheduled for the table.

In conclusion, it must be said that ducks are among the most companionable and interesting of livestock, and have a place on every smallholding. We spent several years selectively breeding and improving our stock of Welsh Harlequins and reaped the benefits of their egg production and supply of roast duckling dinners. Unfortunately, at the time of writing, all our ducks have just been stolen, leaving only the drakes behind. There was no question of it being a fox, for the fox shows no signs of sexual discrimination: all the evidence points to two-legged thieves, particularly as there have already been similar raids in the area.

Geese

Many people who keep geese for the first time are concerned about the provision of a pond for them, but the main need is not for a stretch of water, but for good grazing land. Geese are grazers and grass is their main source of food. They do eat other things, of course, but there is really no point in keeping them unless there is adequate grass. One acre will cater for approximately 9–12 geese, depending on the breed and the quality of the pasture.

The domestic goose has traditionally played a major role in rural life. The frequency with which the name crops up in place-names, testifies to this – 'Goose Lane', 'Goose Green' and 'Goose Drive' are familiar in many areas. Most villages had their stocks of geese which were frequently grazed on the green. When ready to sell, the flock was taken to market in nearby towns, with the drovers often taking several hundred at a time. East Anglia used to supply London with its Christmas geese, and it must have been an amazing sight to see this annual migration of thousands of geese flocking their way along the country lanes and into the heart of the City. They must have walked off a large proportion of fat by the time they arrived. There were also 'goose fairs' in many areas, with the one in Nottingham still surviving as an annual event.

The rise in popularity of the turkey killed off the goose in commercial terms and it remains virtually the only livestock which has not been selectively bred for intensive production. Domestic geese look the same now as they have always done, while other stock has seen dramatic changes. The modern pig, for example, is much longer than it used to be. Sheep are heavier and have shorter legs and the turkey is so overbred to have a large breast and short legs that most of the heavy strains are unable to mate naturally so that artificial insemination has to be used.

There are distinct breeds of geese available, however, and the Embden and Toulouse are the heaviest and most suitable for the table. The Roman is lighter, with distinctive blue eyes, while the Chinese is slighter still. Apart from these there are many anonymous farmyard

Embden geese – one of the best table breeds

Embden × Roman geese in their over-wintering quarters at Broad Leys

white geese whose ancestry is unclear, but which probably form the remnants of the old goose trade of the past. Such geese are referred to as British geese, although this is not recognized as a separate breed. Nevertheless, there are some excellent birds to be found amongst them and many of them are better than some of the more exotic breeds kept for shows.

My family had always kept geese, and Wales retained its tradition of the Christmas goose for longer than England. In fact, I was 21 before I ever tasted turkey. Even after we had moved to live in the north of England, my parents still received a goose from our home village, posted to be received on the day before Christmas Eve.

When we moved to Broad Leys, we decided to buy some breeding geese and to use a proportion of our grazing land for the production of Christmas birds. As it is much more difficult to kill and pluck geese than it is other fowl, we only wanted to retain one for ourselves at Christmas; the others could be sold in the pre-Christmas auction in the nearby market town of Saffron Walden, or to local butchers. Alternatively, there is always a good demand for goslings.

Housing and Management

Adult geese are extremely hardy. They have such an efficient insulation in their layer of down feathers that even the worst of the winter weather does not affect them. Housing need only be very simple, just enough to provide a roof and walls, but in areas where foxes are a problem it may be necessary to have a wire mesh door to lock them up at night. The problem is that geese show no inclination to be housed, preferring to sleep out in all weathers. I must admit that, although we constructed a simple shelter for our geese, we eventually gave up trying to get them to use it and they slept outside. Only on one occasion did a fox try to take one, but although he managed to grab it around the neck, he could not succeed in dragging it through the hedge. The loud cackles of warning alerted us and we dashed out in our night things and wellingtons, releasing the dog as we went. He drove off the fox and we retrieved a shaken and wobbly goose which had lost a lot of its neck feathers, but was otherwise unharmed.

We acquired a breeding trio of one gander and two geese. They were Embden/Roman crosses and our plan was to house them in one section of paddock for over-wintering, and to allow access to the remainder only when the grass was actually growing. In this way, there

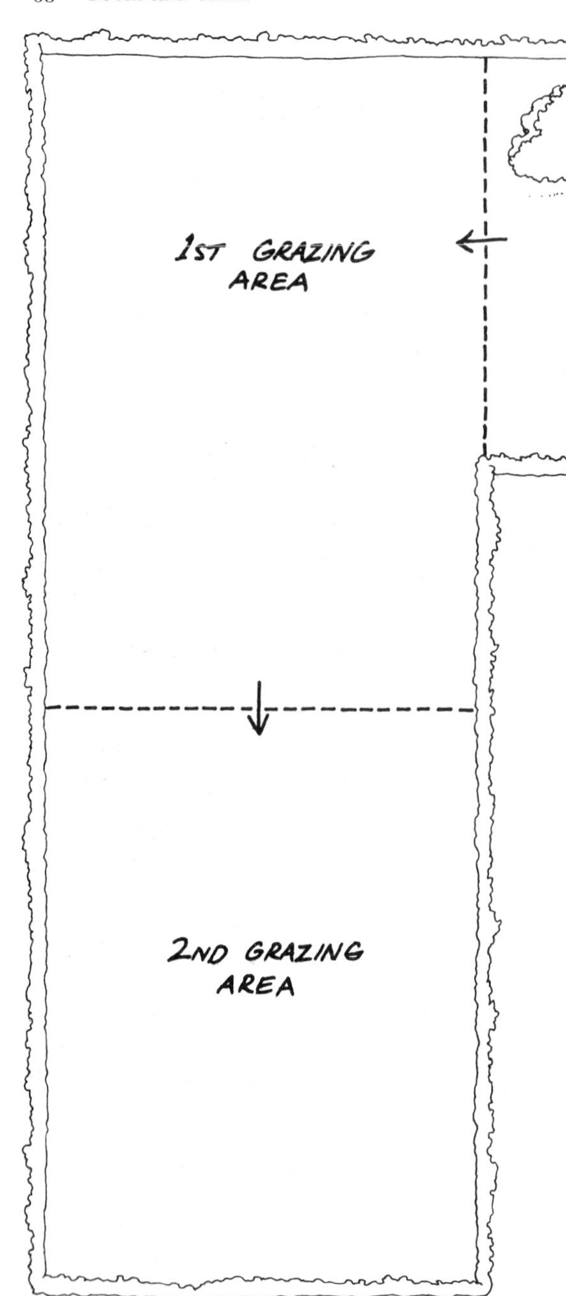

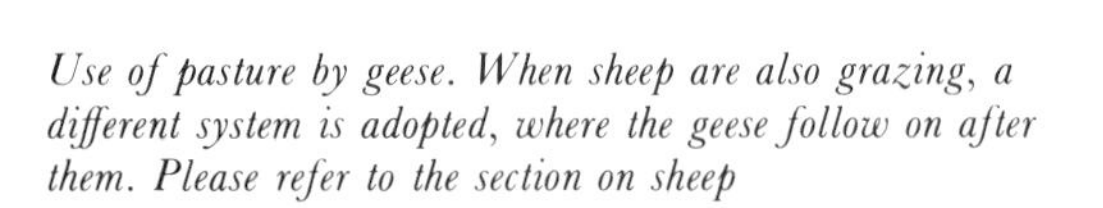

Use of pasture by geese. When sheep are also grazing, a different system is adopted, where the geese follow on after them. Please refer to the section on sheep

would be a rotation in grass use. We constructed the house, but as mentioned earlier, they refused to go near it and we eventually abandoned it. The area of paddock which serves as winter quarters has several large and ancient crab apples. These in themselves provide shelter, but in the late autumn and early winter they also give a rich harvest of scarlet and gold fruit. We use a certain proportion of this for making crab apple jelly and home-made wine, but the bulk of the crop is allowed to fall where it provides food for the geese at a time when the grass has stopped growing and so has little feed value. The geese are not fed exclusively on the fruit, but are given a basic ration of wheat once a day. Many people keep over-wintering and breeding stock in orchards because of the availability of windfalls. We decided against this because the grass in the orchard is kept for the grazing hens and we prefer to keep the various poultry separate in order to minimize disease.

As far as confining geese is concerned, they are normally secure with a 90 cm (3 ft) high fence. As they are heavy, waddling birds they are not able to fly easily over an obstacle. Their tendency is to push under or through a barrier. For this reason, any wire netting used must be well anchored to the ground and have no gaps or weak spots in it. Geese which get into a vegetable garden can cause havoc. They will graze on lettuces and other vegetables and their webbed feet will compact the soil. They will even walk over and break plastic cloches. I have already mentioned the loss of our gooseberry crop one year when the geese got out of the field and ate all the young shoots.

Feeding

From the time in late April and early May when the grass really starts to grow, the adult geese feed exclusively on grass, but as soon as this begins to decline in late July, they are given supplementary rations of wheat. Windfall fruit is a useful stop-gap after the grass. Once the weather becomes really severe it is worth adding a few oats to their wheat ration. Goslings should be given

Over-wintering breeding geese taking advantage of windfall crab apples

proprietary chick crumbs until they are about eight weeks old, and then have that ration gradually reduced as they go over to eating grass. Geese which are being fattened for Christmas can be given a greater concentration of oats in the diet – a 1:1 ratio of wheat and oats is suitable. Barley is also a good addition to the diet if it is available. An expanse of water for swimming is not essential. As with ducks, what is required is a supply of drinking water in a container which is deep enough to allow them to immerse the head and neck and to splash the water over their feathers.

Breeding

Geese are long-lived creatures and it is not uncommon to call at a farm and find that there is a 25-year-old goose there. They are also selective in the choice of mate and a gander can be extremely faithful to one particular goose, while driving off another one. In fact, we once had to sell a goose because she was ostracized by the gander. The best time to acquire new breeding stock is in the autumn so that they have time to sort out their breeding sets and give you the opportunity of spotting any misfits.

Egg laying traditionally starts somewhere around St Valentine's Day, 14 February, and it is not unusual to find the first egg on this very day. This tendency may also have contributed to the folklore associated with goose fidelity. It is a good idea to remove the first eggs, not only to ensure that only definitely fertile eggs are incubated, but also to stimulate the goose to lay more. Once she has a clutch she will become broody and will sit and incubate them. Goose eggs can be used for any purpose where hens' eggs are appropriate; as a general rule, one goose egg is equivalent to three large hen's eggs. They are also much in demand, not only for culinary purposes, but by craftspeople who paint the shells and sell the finished product at craft exhibitions and rural fairs.

Goose eggs can be incubated quite satisfactorily in an incubator and the average incubation period is 35 days. If the geese are hatching the clutch themselves, it is worth providing some kind of large, shallow container of water, in addition to the drinking water supply, so that the sitting geese can dabble in it when they leave the eggs periodically. If the wheat ration is placed near this they will certainly use it. Water is important because it enables them to get the breast and abdomen feathers wet and when they go back to the nests this moisture is transferred to the eggs, ensuring adequate humidity.

Geese need grain to supplement their diet, particularly when the grass declines

This is what happens in nature and it means that the goslings do not become trapped in the shells because inadequate moisture has partially dried the internal membranes. Where a water supply cannot be provided for dabbling it may be necessary to spray the eggs with tepid water, but this is difficult to do because a sitting goose is extremely aggressive. Her calls will also bring the even more aggressive gander to her aid.

Geese are not normally aggressive out of the breeding season, but in the early part of the year the gander, in particular, can be fairly dangerous. For this reason they should be confined to an area where small children cannot gain access. A blow from a gander's wing can break an arm and a pinch from his bill can be painful. A dustbin lid makes an efficient protective shield at feeding times. Where the site needs protection against trespassers, geese make excellent watchdogs and many people, including some business concerns, use them for this purpose.

One of the problems that may be encountered is the goose's choice of nesting site. It is frequently the case that a small house with nesting straw is made available but is ignored by the goose. She will often prefer to make her own out of pieces of twig in a place where there is little protection. In a case like this there is no question of moving her; the best solution is to build a temporary structure around her. Straw bales with an old door placed across the top are most effective.

Another problem that may crop up is the tendency for two geese to want to share the same nesting site. The difficulty here is not associated with the geese themselves, because they are usually happy to do this, but with mixing the two clutches of eggs. If they have been laid at different times and subsequently become mixed up, some will hatch before others. There is the danger that one goose will cease sitting while she still has fertile eggs and these can easily cool and die before they are spotted and placed under the second goose. One solution is to put distinctive marks on the eggs as the clutches are being laid. Before sitting commences, the goose will lay an egg, then cover it up with twigs, straw or whatever nesting material is available. It is only after she has laid a whole clutch in this way that she will start sitting. It is necessary to mark the eggs before this happens because access to them afterwards is difficult.

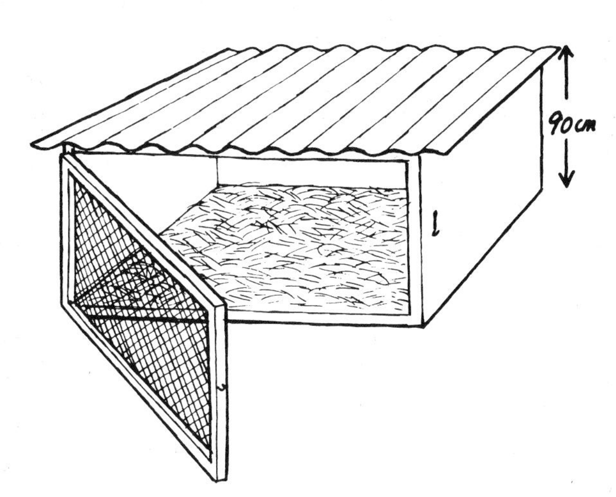

A portable shelter for geese. The length and width of the shelter will be dependent on the number of geese to be housed. Alternatively, several smaller shelters can be constructed which will also serve as broody coops. A gate is only needed if foxes are prevalent

Geese are usually good mothers and, once the goslings have hatched, both the goose and the gander are very protective of them. The main danger at this time is from rats which can kill and drag a gosling away before the parents even realize it is there. It is best to put a goose and goslings in a protected run or, alternatively, to remove the goslings entirely and put them in a protected brooder.

Goslings will do well on chick crumbs and fresh water, and chopped chives and chopped fresh grass are also well received. They grow extremely fast and can either be sold at this age or kept until they are older. Alternatively, they can be reared as table birds for Christmas. If the latter is the case, be certain to place identifying leg rings on the permanent breeding stock because identification may become difficult later on. Poultry equipment suppliers normally stock leg rings.

Goslings need protected conditions until they are sturdy and well-feathered. When they are young they can easily succumb to hypothermia. If this happens, quick action may still save them. Put them immediately in an incubator or airing cupboard, or even a plant propagator, but remember to give them insulation under the feet. If they recover, they will do so fairly quickly. If not, there is nothing else you could have done for them.

The goslings will need to be fed on chick crumbs for about eight weeks, then as new, fresh grass becomes available, the crumbs can gradually be withdrawn. If there is insufficient or poor quality grass they will need supplementary rations in the form of grain.

The Christmas Goose

A mixture of wheat and oats and barley, if available, will provide a good ration for fattening geese and this will need to be given morning and evening until killing time. A goose is more difficult to kill than other poultry and the traditional and accepted way is to lay a broom handle across its neck, hold this down with the feet and then to pull the body up sharply. The neck is quickly broken in this way, but it is easier to carry out with a helper because of the size and weight of the bird. Plucking and gutting is the same procedure as that detailed for other poultry. The Christmas goose has a much better flavour if it is hung, head down, for a few days after killing and before gutting.

Despite the availability of breeding pens, these geese have made a communal nest in the corner of the field

Health

Although goslings are liable to become chilled very easily, once adult they are extremely hardy. The main problem likely to arise is from the over-use of grass. The parasite gizzard-worm can be a nuisance, causing the condition known as 'going-light', a good description of the main symptom. Adult geese can tolerate a certain level of infestation, but young ones are likely to succumb fairly quickly. Watch out for a solitary goose that tends to sit down a lot and is obviously out of condition. A poultry vermifuge given in the drinking water will get rid of the worms and it is advisable to do this once every couple of months while the geese are on grass.

Occasionally, one finds a goose with a 'slipped wing' where muscular weakness results in a wing being dragged or even, in some cases, jutting out at a right angle. There is little that can be done about this and it will not affect table birds, but it is best to avoid using such birds for breeding.

6 Turkeys and Quail

Turkeys

The early American settlers found wild turkeys in abundance in the forests of the New World. The birds became such an important part of their diet that they were hunted almost to extinction in some areas. Fortunately they had been brought back to Britain and other parts of Europe and here, as well as in the Americas, selective breeding resulted in heavier birds for the table.

The original wild turkey was much lighter and slimmer than the examples we see today. It could fly up to perch in the branches of trees for protection at night, a feat which would be beyond the capabilities of the modern turkey. In fact, selective breeding in order to produce broader breasts and shorter legs has been carried out to such an extreme that natural mating is

Turkeys may look like vultures but they make good Christmas dinners

The Broad Leys turkey house, made out of recycled materials

impossible for some heavy strains. Because of this and to increase the overall rate of fertility, artificial insemination is the normal practice in the commercial world.

The original turkeys were bronze in colour and there were several different varieties in different parts of the American continent. Selective breeding produced a wide range of variations but the Broad Breasted Bronze and the Broad Breasted White were two of the most popular. In Britain, before the turkey industry developed to its present size, the Mammoth Bronze, Black Norfolk, White Austrian and Cambridge Bronze were the varieties most commonly kept by farmers. These are still occasionally to be found but are now more likely to be kept by poultry fanciers than by commercial poultry keepers. All commercial turkey strains are now white and tend to be known by rank and number rather than by name. An example of a commercial hybrid strain is Arnewood International Treble CCC. These modern strains are nowhere near as attractive as the older types which had colourful and flamboyant plumage, but they do put on weight much more rapidly and make better table birds.

Housing and Management

Farmers used to keep turkeys outside, free-ranging on grass, but as they are vulnerable to a condition known as Blackhead which tends to be found in damp, waterlogged areas, it was usually the drier areas of the country such as Wiltshire and East Anglia which provided turkeys for market. Now, the practice is to rear them indoors in environmentally controlled conditions.

We decided that we would compromise and keep our turkeys in a barn but that it would also have the benefit of natural light and ventilation. The problem was that we had no barn available so we had to design and build a low-cost turkey house. This was not as difficult as it sounds. Turkeys, like any other fowl, have simple needs: weather and draught protection, dry flooring and good ventilation. We could provide all these things and still use recycled materials for the construction.

We decided to make the turkey house next to a stable so that one wall was immediately available, leaving only three others to worry about. As the area of ground on which the stable stood was a chalk outcrop and naturally free-draining, there was no need to lay a floor; the turkey house would be rather like the traditional American pole-barn, walls and a roof put up on the bare ground.

As the house was to be an extension of the existing stable, the dimensions were tailored to fit. It was 2·4 m (8 ft) high at the front and 2·1 m (7 ft) at the back. This allowed rain to run off the roof towards the back of the building. The house was 3 m (10 ft) wide and 2·4 m (8 ft) deep, which was big enough to house up to a dozen birds. The upright timbers were 10 × 7·5 cm (4 × 3 in.) and were set in holes in the ground to a depth of 60 cm (2 ft). The roof timbers were 10 × 5 cm (4 × 2 in.) and these were nailed on to the uprights. Once the framework was complete, planks were nailed onto the uprights to form boarded walls to a height of 60 cm (2 ft). Corrugated plastic was then fixed onto the area above the wooden planks to complete the walls and to form the roof, but the front south-facing wall where the door was installed was covered in wire mesh. The completed structure was strong, weather-proof, allowed in light and provided adequate ventilation. Its most severe test was in the severe winter of 1981–82 when it stood up to an enormous weight of snow without showing signs of damage. The corrugated plastic which had been recycled from our verandah roof proved to be a most effective roofing material.

The turkey house needs little by way of furniture. A suspended feeder and drinker ensure that food and water are kept clear of litter. When the young turkey poults are first introduced it is important to check that the feeder and drinker are not placed too high for them but as they grow, these will need to be raised gradually. The floor is best covered with a thick layer of wood shavings or chopped straw so that droppings are readily absorbed and conditions underfoot are kept dry.

It has been mentioned that the original wild turkeys were perching birds. Although the modern turkey has been bred to be heavier so that effective flight is difficult, it still retains the instinct to perch at night. For this reason, we felt that it was important to provide our turkey house with low perches, about 60 cm (2 ft) off the ground. These were low enough for easy access and, we felt, respected the birds' natural instincts, although such consideration would be regarded with hilarious incredulity in the commercial world of intensive turkey rearing.

Turkeys are relatively hardy birds once they are over six weeks old, but they are prey to all sorts of ills at the beginning of their lives. They are susceptible to chills, colds and in particular to Blackhead. At one time, this condition led to serious losses, but modern veterinary practice has virtually eliminated it. Vaccines to prevent attack are administered to the newly-hatched poults and antibiotics are added to feedstuffs to provide protection and ensure rapid growth for the remainder of their lives. This practice of doctoring feedstuffs is a controversial one for there are many people who are concerned at the possible long-term effects on those who eat meat from intensively-reared stock which has been treated in this way. There is no doubt that without medication infection would be rife in the intensive commercial units where stock is kept so close together. For the smallholder however, this intensive approach is inappropriate. He is probably rearing turkeys on a small scale for family use and selling the surplus locally. His birds can be kept on a much more natural basis, with no overcrowding and plenty of fresh air and sunlight.

Our practice is a compromise between the two extremes of total environment control and total free-range. We buy turkey poults from a commercial breeder when they are six weeks old. This means that they have received initial medication and have overcome the first

Six-week-old turkey poults on wood shavings litter

obstacle of surviving the early weeks when they are most at risk. When they go into the turkey house they have plenty of room to move about, they have the benefit of the sun coming in through the wire mesh front wall and they have plenty of fresh air to keep them healthy. They do not have access to pasture or general free-ranging conditions because they would risk picking up infection as well as being at the mercy of foxes.

Feeding

As far as feeding is concerned, turkeys on a smallholding do not need to have the medicated rations of commercial turkeys. These are expensive and, although ensuring rapid growth, are not necessary for the smallholder who can resort to a more natural and less expensive feeding method. Our own practice is to give them a small amount of proprietary turkey pellets in the morning and a mixture of grains in the afternoon. In commercial rearing they would not normally have grain and it is important to remember that a small amount of grit is necessary in the diet so that the grains can be digested effectively. The best way of providing this is to supply the birds with a small container of crushed oystershell grit. It will probably only need topping up about once a month and the birds will help themselves as necessary. Crushed oystershell grit is readily available at livestock feed suppliers.

For those who wish to dispense entirely with proprietary pellets, the following ration is suitable for growing turkeys, but it will need to be well mixed and the wheat should ideally be kibbled (coarsely ground) before being added to the mixture: 1 part bran; 2 parts maize meal or oats; 1 part fishmeal or soya extract; 2 parts wheat. It is important not to exceed the fishmeal ration otherwise the meat may develop a fishy taste. For the same reason, it is wise to dispense with fishmeal entirely for the last two weeks before slaughtering takes place. A small, hand-operated grain grinder is useful for kibbling grain and other feedstuffs such as dried beans where small quantities of livestock feed are required at a time. They are readily available, but should always be bolted down to a working surface before use.

The turkeys one week before killing. Note the provision of a suspended feeder and drinker, perches and floor litter

The ration referred to can be given twice or three times a day, whichever is more convenient, or the feed hopper can merely be filled up once a day so that the turkeys help themselves. The latter course is obviously more economic of time. Fresh water should be available at all times, as well as free access to crushed oystershell grit. Failure to provide the latter will result in the birds eating the wood shavings litter, possibly causing blockage of the digestive system.

It is important to remember to approach the turkey house in a quiet, unhurried way so that the birds do not panic and flap all over the place – if they do this they may damage their wings or each other. Any blood on injured wingtips will invite pecking from the other birds.

The best time to obtain turkey poults is from late summer to early autumn. This gives them plenty of time to settle in to their new conditions and allows for a good rate of growth before slaughtering at Christmas. If they are bought too early there is a possibility of too much growth, resulting in birds which are too large for the average-sized family oven. Stags or male turkeys are generally much bigger than hen birds and it may be more appropriate to joint the males for the freezer. When poults are bought, they are usually available as 'A/H stock', meaning 'as hatched'. This means that they will be a mixture of males and females. Commercially, they would be kept together and given similar rations until the age of six or seven weeks and then separated and fed different rations to take into account their different growth rates. For the small turkey producer this difference is not important and there is little to be gained from separating them.

Slaughtering

The method of slaughtering turkeys is essentially the same as that used for chickens but it should be borne in mind that a turkey is much bigger and has strong wings which can inflict damage on the unwary. It is a good idea to have a helper to hold the bird if you are doing the killing. The plucking should be done as soon as possible after death and some people prefer to suspend the bird upside down while this is taking place because of the weight.

We normally raise between six and a dozen birds at a time, depending on how many family orders we have. We allow for a Christmas bird for ourselves, an Easter one and a 'special occasion' one. The latter two are frozen whole. In addition, we joint some of the larger stags and freeze them as portions.

Once the turkeys have gone, the house may be vacant until the following late summer. All the litter is cleared away and put in one of the compost heaps to rot down until it is ready for digging into the vegetable garden. The feeder and drinker are removed, cleaned and stored until needed and the perches are brushed and given a coat of creosote. In the spring we often buy in about a dozen Cobb table chickens as day-olds and for the first few weeks provide them with an infra-red lamp in an indoor room which is rat-proof. Once they are feathered and hardy, they are transferred to the turkey house for growing on until they reach table weight. When they have been slaughtered, there is still plenty of time to clear out and disinfect the house before the next consignment of turkey poults arrive. Further details of our way of rearing table chickens are given on page 57.

Quail are unobtrusive but highly productive little birds

Quail

The quail is an extraordinary little bird. About a third of the size of a normal hen, it has the distinction of being California's national bird, as well as producing eggs which are regarded by connoisseurs as one of life's gastronomic delights. There are several breeds, some such as the painted varieties being regarded as 'fancy' breeds and kept for show. The main domestic breeds for the production of eggs and table birds are the Japanese or *Coturnix japonica* and the Bob White or *Colinus virgianus*. Both of these have been developed to produce commercial strains with some being bred for egg production, while others have been selected for increased weight. However, the distinction is nowhere as near as extreme as it is with chickens, and, for the purposes of the smallholder, the strains may be regarded as dual purpose.

Our particular interest in quail arose out of correspondence within the magazine. There appeared to be little information available about these fascinating little creatures, so we decided to embark on a project that would enable us to discover at first hand what the advantages and possible snags were. We made contact with a breeder of commercial strains and ordered a breeding trio of one cock and two hens. They were despatched by rail, to be collected at the local station. When we opened up the travelling box, we found three sprightly and bright-eyed little birds, and two blue and brown patterned eggs. It was a hopeful beginning.

Housing and Management

The chief enemy of the quail in Britain is the rat. It will attack, kill and carry off a full-grown quail with ease, and its cunning and expertise in penetrating defences should not be underestimated. We had prepared what we thought to be rat-proof housing, a small poultry house in a completely enclosed run. The turf was stripped off a section of the orchard and several thicknesses of close-meshed wire netting placed on the soil. The turf was then replaced and 1 metre (3 ft) high close-meshed walls attached to wooden stakes set into the ground were attached to one side of the house. The run was covered over with the same type of netting so that foxes would be kept out as well. Access to the house for feeding and watering was via the door, while the birds could get in and out through a pop-hole. This was closed in the evenings, as an added precaution in case a predator did manage to gain access to the run. The house was equipped with a suspended feeder and drinker, a perch and two nestboxes which could be opened from the outside for egg collection.

The quail settled down and continued laying at such a pace that a surplus of eggs soon built up. Some of these were used for cooking, while others were set aside for incubation in the incubator. Neither of the hens showed any inclination to become broody. As it happens they are unreliable in this respect and much better results are obtained by artificial incubation. The birds are extremely quiet and unobtrusive, making them, with rabbits, the ideal livestock for the town where neighbour complaints in relation to noise are fairly common. They produce a range of subdued chirrups and warbles, but no loud cackling or crowing as one finds with chickens.

Feeding
Proprietary chick crumbs are an ideal food for quail of all ages, with budgerigar seed providing the grain requirements. They also appreciate a few shredded lettuce leaves and water should be available at all times. Grit will be needed to help in the digestion of grain and this should be provided in a small separate container from which they can help themselves as required. Our practice is to feed them morning and evening as part of the general livestock round, and the easiest method is merely to check and, if necessary, top up their supplies of food and water. The eggs are collected at the same time.

This system worked well for quite a long time, but one night despite all our seemingly thorough preparations, the whole lot were taken by rats. We had neglected to close the pop-hole, and by incredible bad luck, this coincided with a rat breakthrough into the run. They had obviously found a weak point in the underground netting, gnawed through it, emerged in the run and then found that the pop-hole had been conveniently left open for them. All that was left was a feather or two. Fortunately, we had fertile eggs incubating and so could replace the stock, but this time we decided to play safe and use rabbit hutches which stood well clear of the ground as future housing.

Breeding
The incubation period of quail is 18 days and proceeds in exactly the same way as for other poultry. The eggs need frequent turning and the water level of the incubator should be kept topped up so that the humidity is at the appropriate level. With such a short incubation, it is not worth trying to candle the eggs, and, anyway, the dappled pattern on the shell makes fertile eggs extremely difficult to spot.

When the eggs hatch, the baby quail are extremely small, not much bigger than bumble bees. Brooding them once they are taken out of the incubator presents problems of confinement because they can escape through the tiniest of crevices. We found the best solution to be a large fish tank with an infra-red lamp suspended above it. They were secure within its solid walls, with the whole unit safely installed in the conservatory.

The chicks will soon start eating chick crumbs and for the first few days these need to be crushed to make the particles smaller. After about a week, this will no longer be necessary. Care needs to be taken in the provision of water while they are tiny: even the smallest of chick drinkers is unsuitable at this stage because they can get into the water and possibly drown. We found the best solution to be a small shallow container with pebbles placed in it so that, while water was available, it was too shallow to drown in.

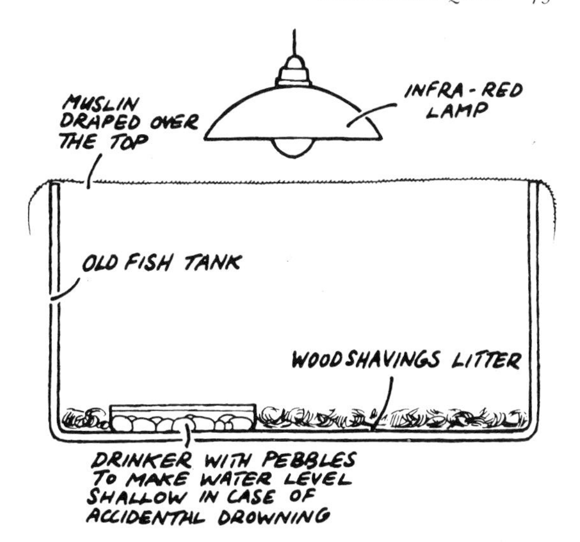

The safest way we know of brooding baby quail

The infra-red lamp was gradually raised as they became more hardy until it was dispensed with entirely. They were kept in the fish tank until they were big enough to go into the rabbit hutch, and were sufficiently hardy to cope with the night-time temperatures.

The Quail Harvest
The eggs can be used for any purpose where a chicken's egg is called for. Approximately three quails' eggs equal one hen's egg. They can also be hard-boiled and preserved, but care must be taken not to boil them for more than 30 seconds, otherwise the shells are hard to remove. Once they are shelled, they can be pickled in brine or in boiled and cooled, spiced vinegar. Alternatively, they can be preserved in aspic to produce the well-known delicacy.

Quail also make delicious roasting fowl. They are killed, plucked and gutted in the same way as other poultry, wrapped in bacon and roasted in a medium oven. One to two quails per person will be needed and the young are ready for slaughtering from the age of five weeks onwards. They also freeze well.

Health
Quail are subject to some of the ills that affect other poultry, but are unlikely to contract anything serious if they are kept away from other birds. Apart from the obvious precautions of regular clearing out of droppings, cleaning feeders and drinkers and giving adequate food and clean water, there is little else to worry about. They should, of course, be regularly examined for any evidence of lice and mite attack and given an occasional dusting with a proprietary powder.

7 Rabbits and Bees

Rabbits

In the past, wild rabbits were often all that stood between the peasantry and starvation. Even in the period immediately after World War II, smallholders gained a useful income from them. One of my uncles used to catch rabbits and be paid a bounty for each tail that he produced; the rest of the rabbits made tasty pies which I still remember with great affection. The disease myxomatosis stopped all that and from the early 1950s, there was rapid decline in the numbers. At the same time, there was a public reaction against rabbit meat which is still evident today. Although the numbers of wild rabbits are again on the increase, there is still an attitude that they are not safe to eat. This is borne out by the fact that rabbit now sold in the supermarket bears the label either 'Chinese', indicating its country of origin, or 'Tame rabbit', to show that it is from a commercial rabbit farm and not a wild warren.

There is a considerable difference in taste between the meat of a wild rabbit and that of a domestic one. The former may have a stronger 'gamey' taste and needs to be soaked in water with a little vinegar for a few hours before cooking, in order to tone down the flavour. The meat of domestic rabbits is much lighter in colour, more delicate and resembles chicken in taste and texture.

Rabbit Breeds

There are many breeds of domestic rabbits, including a great number which have been developed for show appearance or for the quality of their fur. For the table, however, there are really only two breeds to consider, the Californian and the New Zealand White. Both were bred and developed in the USA and introduced to

Young New Zealand White rabbit

Britain in the 1940s and 1950s. They have a much faster rate of growth than other breeds and have more meat in relation to bone. The New Zealand White is an albino and is consequently pure white with pink eyes. It grows at a slightly faster rate than the Californian and is, therefore, the one most often kept by commercial rabbit farmers. The Californian, although slower-growing, attains a heavier weight and has a more attractive and substantial pelt. The coat is white with black or dark brown on the nose and tips of the ears. Both these breeds are available from commercial breeders and it is wise to buy only accredited stock which is guaranteed healthy.

Housing and Management

Pet rabbits are normally kept in hutches, while commercial ones are in cages which allow the droppings to fall through. There is less work involved in cleaning in this way, but the wire floors of the cages are less comfortable for the rabbits' feet than hutch floors.

We decided that we would house rabbits in such a way that humane considerations would be respected, yet convenience and labour-saving factors would be taken into consideration. We put up commercial cages in an outhouse and placed a section of wooden flooring on one side of each cage so that, in each case, the other side remained as a wire mesh floor. This compromise means that droppings and urine fall through to the floor below, where they are absorbed by straw which is periodically removed. The rabbits soon learnt not to foul the wooden side. During periods of fine weather, the rabbits are put in temporary grazing arks so that they have the benefit of grazing on grass. Only females or a mother with young can be housed together in this way. The buck needs to have his own quarters, otherwise indiscriminate and possibly harmful breeding will take place, as well as fighting.

This system works well for us. Our basic breeding stock is one buck and two does; each has a cage housed in an outhouse which has electric light, and which provides a protected environment for hard winters. They spend the nights there, as well as periods of wet or severe weather, but have the benefit of grazing in fine weather. Each cage has a feed hopper clipped onto it so that food can be given from the outside without having to open up the cage. There is also a hayrack for the provision of hay and greens and a bottle drinker fitted with a tube attachment. The rabbits soon learn to drink from the tube and it is much more hygienic than some of the more traditional drinkers where droppings could get in to contaminate the water. The grazing arks are fitted with drinkers only, for while the rabbits are outside during the day they eat only grass, having had concentrates in the morning.

The arks are moved on periodically, so that fresh grass is made available and there is a shelter at one end into

Rabbits grazing on the lawn in a home-made grazing ark at Broad Leys

A different design of ark used by a mother and her litter

which the rabbits can run if there is a sudden shower of rain.

Feeding

Rabbits will eat a wide variety of greens and vegetables from the garden and these undoubtedly are useful sources of food for the smallholder's stock. They do need concentrates as well, if they are to thrive and make good table rabbits. Proprietary rabbit pellets are available from livestock feed suppliers, but they are expensive. They also contain antibiotics to counteract coccidiosis which affects rabbits as well as poultry. Commercial rabbits are fed exclusively on these pellets, but many smallholders prefer either to give their rabbits a few pellets in the morning and feed hay, greens and oats the rest of the time, or to forgo pellets entirely. It is a personal decision. Rabbits undoubtedly fatten more effectively on an exclusive diet of proprietary pellets, but the smallholder who is raising them for his own use will not need to achieve crucial feed conversion ratios like the commercial supplier. We prefer to give a more natural, varied and cheaper diet and have rabbits which take longer to achieve table weight, but which produce good quality meat untainted by antibiotics.

The average daily ration would consist of 113 g (4 oz) of pellets a day, or approximately 195 g (7 oz) of other foods, plus hay. If other food is given, it should have a basic amount of oats and a mixture of chopped vegetables and scraps such as baked and crisped potato peelings or toasted crusts. Avoid too much of any single item; then there will be less likelihood of digestive upsets.

Breeding

Breeding will take place all the year round, but only if there is artificial light provided in the winter. Even so, winter breeding can be difficult and I would not recommend it unless it is vital to have non-stop production. If the rabbits are for your own freezer, twice or three times a year is probably adequate. One buck and two does of the New Zealand White breed are capable of producing up to 50 rabbits a year, which is more than enough for the average family. We find it enough to breed them twice a year and have some for the freezer and some for sale.

For breeding purposes the doe is put into the buck's cage; never the other way around otherwise the territorial doe might attack and injure him. The correct way to lift a rabbit is to hold it by the scruff of the neck and to support its bottom. The ears are delicate and should never be used to lift a rabbit. Mating will normally take place fairly soon, but if the doe refuses to stand still, it may be a good idea of follow the commercial practice of 'assisted mating'. This involves grasping the doe firmly by the scruff of the neck and placing the other hand under her belly so that her hocks are lifted slightly off the ground. In this way the buck has easier access and mating can take place more quickly.

Once the doe is back in her own quarters, a record should be kept of when mating took place so that adequate preparations can take place before the birth of the young. The average gestation period is 31 days and during this time the doe should have normal rations, without too many starchy foods which will tend to make her too fat. During the last week, she will need extra rations to cater for the increased demands on her system and to prepare for lactation. Raspberry leaves are a useful addition to the diet at this time for they have a beneficial effect on the uterine muscles before the birth takes place. (Raspberry tea was traditionally drunk by pregnant women for the same reason.)

From the 27th day, it is wise to introduce a nestbox if cages are used. This can be made of wood or plastic; a large plastic cake storage container is ideal. This is half-filled with soft hay and placed in the protected side of the cage or hutch. The doe will pluck fur from her own body and line the nest with it. Once she starts doing this, it is a sign that the birth is imminent. She is best left alone while giving birth and until she has cleaned and covered over the young ones with fur in the nest. In fact, the first indication that anything has happened is usually when you suddenly notice a heaving mass of soft fur and hay in the nestbox. Closer examination will reveal several pink and hairless babies with their eyes closed. It is important to check for still-born or defective ones, which should be removed. If you have previously established a good relationship with the mother, she will not object to this. In fact, I have never found any evidence to support the old claim that mother rabbits will eat their young if disturbed, although it could be a more common practice in the wild.

The mother will periodically hop into the box to feed the little ones and will continue to do so for several weeks. While they are tiny, she will carefully cover them over each time, to keep them warm until their fur grows. As they grow, they become more adventurous, and will eventually not use the nestbox at all. At this stage it can be discarded and, as long as the weather is mild and sunny, they can accompany the mother when she is put out in the grazing ark.

The doe will continue to feed them until they are about five weeks old, when they will be weaned. Before this, they will already have begun to gnaw solid food, and it will be necessary to provide extra rations to cater for them. In commercial rabbitries, the practice is to increase the doe's basic ration of 113 g (4 oz) pellets to 225 g (8 oz) a day, a week before the birth, to cater for increased bodily needs and to continue this ration while the doe is lactating. Once the young begin to show an interest in solids, the pellet ration is given on an *ad lib* basis; in other words, the feed hopper is kept topped up so that they can help themselves as necessary.

Now is a good time to sell any stock which is not destined for the freezer, and a notice outside the gate or advertisement in the local paper will frequently bring buyers. The local pet shop or feed supplier is also a good place to put up a notice. There is usually no charge for this. Local self-sufficiency groups frequently publish a newsletter for their local members, as well as conducting regular meetings. Here there are excellent opportunities for selling or bartering surplus stock.

Slaughtering

Rabbits which are to be killed for meat should be separated from the doe. We find it convenient to keep them as a small colony in one of the arks and at this stage, to keep them outside. This is another good reason for breeding rabbits in the summer months only. The killing time for a non-commercial smallholder is not as crucial as it is for a commercial rabbit producer who aims to have 2–2·2 kg ($4\frac{1}{2}$–5 lb) rabbits in 9–10 weeks. If they are kept longer than this, it will be necessary to separate the sexes, otherwise early breeding may result. It is much more convenient to kill them at this age, even though they may not be as heavy as commercially reared ones. Food should be withheld from rabbits due to be killed for a period of 12 hours beforehand, but they should have water. The simplest and most effective way of killing is to hit them on the back of the neck with the so-called 'rabbit punch'. This is administering a chopping blow with the side of the hand or blunt instrument and can either be done while the rabbit is on a horizontal surface, or when held by the back legs so that it is suspended upside down. In the first case, it will be necessary to hold the ears up and slightly forward to give clear access to the back of the neck and head. Alternatively, death can be by neck dislocation and is achieved by a method similar to that adopted for poultry. Many people prefer to hit the back of the neck first and then dislocate it. It should be emphasized again that no one should attempt to do it without having had the technique demonstrated, for death must be instantaneous.

The skin is easy to remove while the carcass is still

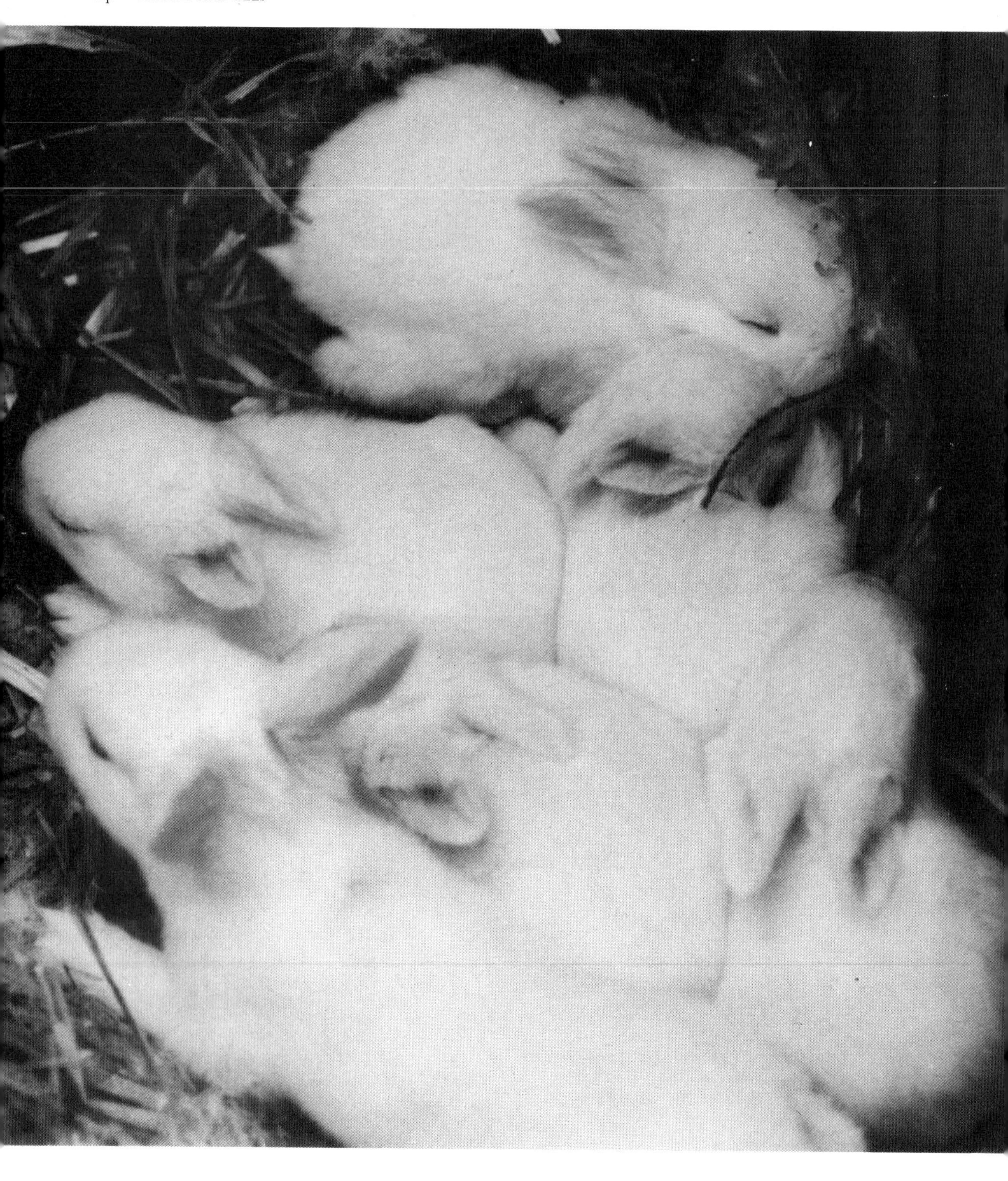

Two-week-old New Zealand White rabbits

Meat rabbits awaiting slaughter

warm. Pinch the skin of the belly and pull it forward slightly in order to make an incision with a sharp, pointed knife which will not pierce the body wall beneath. Extent the incision upwards to the neck and downwards to the vent. Cut the skin around the neck and back of the head and around the tail and front and back legs. Now make cuts at an angle from the main incision along the front and back legs in order to join up with the circular cuts. Pull the skin away from the body wall, using one hand to break the connective tissue holding it, and it will come away in one piece. Some people prefer not to make a central cut, so that the skin can be pulled off in one piece, like a jersey. It is then stretched on wire to dry.

To gut the rabbit, make an incision in the body wall on the belly side and establish where the gut lies underneath, so that the knife point does not pierce it. Extend the cut to the breast bone and down to the vent, so that all the guts are revealed. They can now be taken out quite easily, but take care not to puncture the lower bowel and gall bladder which is attached to the liver. The kidneys can be left in or removed, as desired. Cut off the head and legs after first snapping the latter against the edge of a table. The carcass is now ready for roasting or freezing whole. Alternatively, it can be jointed.

Rabbit skins can be cured quite easily. This can either be done as soon as gutting and other operations have taken place or, alternatively, the skins can be temporarily stored until required. They should be rubbed with salt and left to dry in the sun until an opportunity arises for curing them. This process is described on page 123.

Health

Few problems should arise as long as the rabbits are properly housed, fed and watered, droppings are regularly cleared away and feeders and drinkers are kept clean. Mention has already been made of the wisdom of acquiring the original breeding stock from an accredited breeder whose rabbits are guaranteed healthy and free of disease. It is a good idea to have breeding stock injected against myxomatosis. A commercial breeder will often do this for you, as an optional extra, or a veterinary surgeon will be able to help.

A regular check should be kept on ears and feet. Mites in the ear can be troublesome and are indicated by excessive twitching of the ears and rubbing them with the paws. An examination will reveal orange-coloured secretions in the ear passages. These should be carefully removed with cotton buds dipped in warm, soapy water and a few drops of veterinary canker liquid administered. It is a good idea to put some more drops in the ears a week later to ensure that all the mites, plus any that might have subsequently hatched, have been destroyed.

The claws will need regular clipping for as the rabbit is not burrowing, they become long and overgrown. If the foot is squeezed slightly, the claws will protrude and the position of the blood vessel in each one can be established by looking at it in a strong light. Clip the claw back, but ensure that the cut is well clear of the blood vessel. Ordinary manicure clippers are quite satisfactory.

Rabbits, particularly pink-eyed New Zealand Whites, may suffer eye irritation if droppings and urine are not cleared away and there is a build-up of ammonia. Straw placed under the flooring will catch, absorb and deodorize droppings, but it must be cleared regularly. Hutches obviously need even more frequent cleaning, ideally every day.

Eye watering and a runny nose may mean that a rabbit has a cold – usually referred to as 'snuffles'. A little onion and garlic added to the food will help, but there must also be freedom from draughts and damp, and good ventilation. Coccidiosis is now no longer a problem, mainly because coccidiostat antibiotics are included in the proprietary feeds. It is a protozoan infection, normally picked up from damp litter, so it is vital to ensure that a build-up of such material does not occur in hutches. Where cages are used, the chances of contracting coccidiosis are remote.

Finally, where garden vegetables and green foods are given, ensure that they are clean and fresh, with no hint of mouldiness. Wild foods such as chickweed and Shepherd's Purse can also be fed, but use a good text book of wild plants to make sure you can distinguish between edible and toxic ones. No picking should take place from an area where heavy concentrations of lead from car exhaust fumes are likely, or where dogs have been fouling. Too much of any one vegetable should be avoided in case digestive upsets result. An excess of cabbage can lead to a build-up of gas causing congestion – this condition is known as 'blows'.

Bees

Bees were in the past an integral part of many traditional smallholdings. They were kept, not only for the honey harvest, but also for the pollination which they brought about in the fruit and vegetable blossoms. Many cottage gardens contained also a wide range of flowers which were planted and sown to provide extra forage for bees over as long a seasonal span as possible. Some of these, such as mignonette with its insignificant flowers yet beautiful smell are not as commonly seen in the flower gardens of today, although there is currently a revival of interest in the cottage garden tradition.

Bees were also kept simply because they are immensely interesting insects, with a well-ordered social hierarchy, where each group has its own particular function in relation to the whole hive colony. The queen, the largest insect in the group, is the heart of the colony and lays eggs which provide replacements. The drones are smaller and numerous. They are males whose function it is to fertilize the queen during her so-called 'nuptial flight'. The smallest and most numerous bees are the workers whose job it is to fly to and from the hive foraging for food in the surrounding area.

My family had always kept bees. One of my mother's tasks, as a small child, was to run after the bees if they swarmed in summer. She had to keep them in view until they had settled, hanging like a great brown ball on a tree branch or cottage eave, then she would race home to tell my grandfather where they were so that he could retrieve them.

One of my brothers had kept on the family tradition, so it was natural, when we decided to have bees at Broad Leys, that we should go to him for advice and information. This is essential for anyone new to beekeeping. There is nothing better than direct, personal experience as a basis for making a final decision to go ahead with a project. This is where local beekeeping societies are so valuable, for there are always experienced beekeepers who are ready to help those new to the field. In addition, many local authorities and agricultural colleges run beekeeping courses and it is an excellent idea to take advantage of one of these. This chapter can only be regarded as the briefest of introductions to the subject.

The Hive and Other Equipment

There are several different types of hive, all of which have their own modifications and advantages. They include WBC, National, Commercial, Dadant, Smith and Langstroth. Traditionally the most common was the WBC, a solid and picturesque hive which is still frequently used by small beekeepers and which many claim is the most sturdy in winter. It has been usurped, in recent years, by the National which is now the most common hive in Britain. British standard frames will fit both National and WBC hives, so the small beekeeper will find either of these two to be the most appropriate. Our hives are WBC ones, but this is because we obtained them secondhand from my brother and it was obviously cheaper to use them than to buy new ones. If it had been necessary to buy new stock, we would probably have chosen the National hive.

We sited the hives in the paddock, with their backs to the lane, so that passers-by would not be stung. They were south-facing, with trees behind and to the left of them, giving wind protection and sheltered conditions. In this way, the flight path of the bees took them onto the smallholding site itself, with the vegetable garden and orchard ahead and to the right. A short section of post-and-netting fence kept any grazing animals at a distance, so that they would not be stung and also to ensure that the hives were not knocked over. In fact, the only time this barrier proved ineffective was when the local hunt went up the lane and the fox veered off across our land. All the hounds followed, streaming over the netting and around the hives and putting the geese to flight. By

Interior of a National hive showing a section super with 32 sections (Robert Lee Bee Supplies Ltd)

some miracle, the hives were not overturned, but this escape was in no way due to the efforts of the hunt.

Protective clothing is vital for anyone handling bees, with trousers tucked into socks and gloves pulled over sleeves. A hat with a veil, giving protection to the face and neck, is also essential. There are some experienced beekeepers who are so immune to stings that they handle the bees without this kind of protection, but this procedure is not recommended for a beginner. There is a small minority of people who are allergic to bee stings and it would obviously be inappropriate for them to keep bees at all. The best bee outfit I have ever come across is a lightweight radiation suit. The only problem is in explaining to unexpected visitors that I am really checking the bees and not morbidly preparing for a nuclear holocaust.

A smoker is essential for calming the bees during hive examination, but it needs to be an efficient one, able to produce a puff of smoke at the right time. There is nothing worse than having one which does not work properly, and it is worth buying a new one if there is any doubt about this. A hive tool is the only other piece of equipment which is necessary. This is a purpose-made tool for levering apart frames which may have become stuck with wax.

Taking Delivery of the Bees

The hive should be got ready before the bees arrive. This will entail proper siting and checking that the brood chamber is placed on top of a floorboard and that there is a metal queen excluder placed on top. The brood chamber is where eggs will be laid by the queen, but she will be denied access to the super or honeycomb chamber above by the fact that the hole in the excluder is too small to allow access to any bees apart from the workers.

When the bees arrive from the supplier, they will be in a travelling box. Open the hive and light the smoker before donning the hat and veil and opening the box. Each frame is then transferred from the travelling container and placed, in the same position, resting on the ledges and hanging vertically in the brood chamber. There should be brood combs with eggs and grubs on the frames; these will form the nucleus of the new colony. There will also be bees on the frames and any which remain in the box after the frames have been transferred should be shaken onto the top of the brood chamber. Any gaps in the brood box should be filled with new brood combs with wax foundation on them so that, as the bees multiply, there is room for them to spread. Put the cover on top of the queen excluder and then replace the roof. Some people prefer to put a super, or honey chamber, on top of the queen excluder at this stage. This is a shallower chamber than the brood box and holds the combs on which the worker bees will manufacture and store honey. The queen is excluded from this to prevent her laying eggs there. Some people prefer to put the super in position later, but late spring should be regarded as the latest possible time.

Where a swarm has to be taken, the procedure is slightly different. Swarming takes place for a number of reasons – because the hive is overpopulated, because the weather or forage conditions are not ideal or sometimes

Protective clothing is essential when dealing with bees (Robert Lee Bee Supplies Ltd)

because a particular strain of bee has a genetic tendency towards swarming. Whatever the reason, a new hive should be made ready as before, and the swarm taken from its settling place. This really needs two people and involves one person sawing through the branch on which the bees have landed, or knocking it off, while the second person stands poised below with a box to take the swarm when it comes down. The first time we had to do this we were rather nervous, but by one of those occasional strokes of good fortune, a complete stranger appeared at the gate, enquiring if she could help. She turned out to be an expert beekeeper.

Once the swarm is taken, the box is placed on its side on a white sheet, with the other end leading to the alighting board at the entrance to the brood chamber. The bees will then crawl up the sheet and take residence in the hive. Foundation combs with wax should be made available to them so that they can then start to produce new brood. It will also be necessary to feed them for they will not have their own stores. This is simply a matter of making up a sugar solution in the proportions of 90 g sugar to ½ litre of boiling water (2 lb to 1 pint) and then letting it become cold before pouring it into a purpose-made feeder placed on top of the feeding hole in the crown board or inner cover. It should be kept topped up until it is obvious that the bees are no longer interested because they are getting enough food elsewhere.

Taking the Honey

Honey is ready for extraction when the cells in the supers have been capped by the bees. Until this happens, the honey will not be in a ripe condition and will not store. A few days before extracting, lift off the super and place an escape board between it and the brood chamber. This will allow bees to go down into the brood area, but not back into the super, so that in a few days, there will only be a few bees left to contend with when the honey combs are removed. When the combs are taken out they should be replaced with new wax foundation combs so that the bees are encouraged to make more honey.

The easiest way to carry the frames is to put them in an empty super, and it is much easier if two people carry it because the weight is considerable. The first time we did this we were rather rough and knocked off some of the cappings so that a trail of honey oozed out as we were carrying it. Taking honey is a sticky business at the best of times, but careful handling does minimize this.

Once inside, in a room where angry bees and hungry wasps cannot follow, the frames are uncapped. This is achieved by standing each frame on a tray and slicing off the caps, on each side, with a sharp knife dipped in hot water. For efficient separation, a centrifugal separator is recommended. These are expensive to buy, but are justified if the beekeeping enterprise is big enough. Alternatively, they can sometimes be hired from local beekeeping societies. Once separated, the honey should be left to settle in a covered container so that air bubbles have a chance to escape, and then put in jars. These can either be bought direct from beekeeping equipment suppliers or purchased via a local society. Surplus wax should be rendered down in warm water for reuse.

Once the honey has been taken for that season, the winter store of the bee colony no longer exists. It will need to be replaced by sugar solution or the bees will die. It is our practice to leave a proportion of the hard-earned honey behind, rather than depriving them of it all. They need to have sugar solution as well, but we feel that it is better to work with nature, instead of merely exploiting it. Further details of seasonal management of the bees are given on page 129.

8 Dairy Goats

There are several positive advantages in keeping goats. They browse on broad-leaved weeds as well as grasses and, in this respect, have a general improving effect on pasture. They can be kept in a comparatively restricted area and do not necessarily need grazing land at all. There are many examples of stall-fed goats which have an attached exercise yard and which rely on the owner to bring green food such as hedge trimmings and other browse material. Goats' milk is increasingly regarded as being beneficial to health and there is considerable evidence to indicate its usefulness in cases of allergy such as eczema and asthma. Surplus milk can be sold and there is no legislation in Britain such as that which prohibits the sale of cows' milk by non-registered producers. There are disadvantages, of course, and one of the main ones is the destructive capabilities of goats in relation to trees and shrubs. They will eat their way through the thorniest of hedges and can wreak havoc in an orchard. Electric fencing will confine them, but, where effective fencing is too expensive, tethering may be the last resort.

The Thear family with some of their Anglo-Nubian goatlings in winter 1975

A British Toggenburg goat with Anglo-Nubian cross kids

We decided to keep goats for two reasons. With a total area of two acres, including the house and gardens, there was insufficient room for a cow. The second reason was that our daughter Helen developed eczema; a condition which quickly cleared up when she drank goats' milk instead of that from cows.

There are several breeds of goats in Britain, but the main dairy breeds are British Saanen, British Toggenburg, British Alpine and Anglo-Nubian. They all have their advantages, and it is impossible to be dogmatic and say that one breed is better than another. What is important is the particular 'strain' and whether or not it comes from registered, milk-recorded stock. The British Goat Society is the body which co-ordinates goat-keeping activities in Britain and will put people in touch with goat-breeders in their particular locality. The local goat societies, which are affiliated to the British Goat Society, are also excellent sources of help and advice.

Our first goats were Anglo-Nubians which have distinctive Roman noses and long, droopy ears. This breed has been referred to as the 'Jersey' of the goat world – a reference to the high butterfat content of the milk – but overall yields tend to be lower than those of the other dairy breeds. We eventually sold our Nubians and acquired our present milkers, a British Saanen called Florence and a British Toggenburg named Bilberry.

Housing and Management

There are many way of housing and managing dairy goats, depending on the numbers kept, available buildings and personal choice. The needs, however, do not vary and may be listed as follows: sleeping area, milking area and recreation area. As it is important to cater adequately for these needs, it is worth examining them in greater detail.

Sleeping area This may be regarded as the goat's 'home', for it not only sleeps here, but probably spends a considerable amount of other time as well, particularly in periods of severe weather. Some goat-keepers have communal sleeping areas for their stock, but they are in the minority and most people find that separate stalls for individual animals are easier to manage. Goats are extremely sociable animals so they do not like to be totally separated, preferring to see other goats from their stalls.

The size of individual pens will vary, depending on the overall space available in the goat house, but it should be large enough to allow the goat to move around in comfort. It will also need to be equipped with a hayrack to hold hay and green food, a water bucket, securely supported, and a mineral lick. The floor will need a layer of straw or wood shavings to absorb droppings and provide a warm, insulated surface on which to rest. The litter needs periodic removal, but only after there is a noticeable build-up. Adding a fresh layer of litter on top of the existing layer can take place about once a week, so that complete removal need only take place once every couple of months. Used litter is excellent for composting or as a mulch around fruit bushes.

Milking area It is important that a dairy animal is milked in an area separate from that where it sleeps. It not only makes the activity easier to organize, but also ensures that hygienic considerations are met. Ideally, a milking area should be light, airy and easy to hose down. As far as goats are concerned, a milking stand is useful

Gwilym Thear with Bilberry, a pure-bred registered British Toggenburg who looks just like a British Alpine

because, as they are smaller than cows, a strain is imposed on the back of the milker if he or she has to bend down at an uncomfortable angle. The only other necessity is either a small hayrack or feed bucket so that the goat will have food with which to occupy herself while being milked.

Recreation area This is quite simply an area where goats can stretch their legs, enjoy some fresh air and have a change of surroundings. Where goats are kept in confined areas and are 'stall fed' rather than having access to grazing it is particularly important to have an exercise yard or at least somewhere where they can be taken for regular exercise. Even a daily walk on a lead is better than nothing. Where there is plenty of land, the goats will have enough recreation while they are out browsing and grazing.

As far as our goats are concerned, we worked out a system which catered for all these needs and saved us as much time as possible. We converted part of the range of outbuildings at the bottom of the garden in such a way that what were originally two interconnected buildings became two pens, a passageway, feed store and milking parlour.

Bilberry, our British Toggenburg, when she was a goatling, in the goat enclosure (Stephen Austin Newspapers Ltd)

The two buildings had both been used for horses, but when they were cleared out, we discovered that the smaller of the two had a concrete floor with a drain. This meant that it could be hosed down with water and was, therefore, suitable as a milking parlour. The window space was covered with a metal grille so that ventilation was adequate, but birds were excluded. A milking stand was constructed on one side in the form of a wooden platform which stands 45 cm (18 in.) off the floor. At one end of this is a feed bucket in a fixed metal stand. There is a short length of tethering chain attached to the wall nearby. The only other adaptation carried out was giving the walls and ceiling a coat of white gloss paint so that the room was lighter and capable of being washed down without flaking particles coming off the walls. Electric light in the form of a single light bulb was already installed.

The second building which led off from the milking parlour had two goat pens constructed in it. These were made with wooden walls and metal grilles so that the goats could see out and were, therefore, not isolated from each other. Each pen is equipped with a wooden hayrack, a water bucket in a metal support attached to the wall, and a suspended mineral lick. The effect of constructing these pens was to form a passageway between them and the outside wall. This meant that the goats could be taken for milking without needing to go outside. The remaining space was used as a feed store with metal bins which stored all the livestock feeds. Electric lighting is available in these areas, as well as the milking parlour.

To have constructed original buildings would have cost a fortune, which is why I stressed earlier that anyone thinking of buying a smallholding should make sure that it has existing outbuildings. Most buildings, no matter how dilapidated, are capable of renovation and adaptation without too much expense.

As far as recreation and exercise are concerned, our goats have a small corral with a weather shelter in one corner. The whole area is enclosed by a post and rail and netting fence and there is a gate for access. They are put in here during the day so that they can run around unimpeded. There is no grazing or browsing in the corral, but a large, metal hayrack on the outside wall of the shelter is periodically filled with browsing material. When the grass is actively growing they are tethered outside and allowed to graze, usually in the paddock, unless it is a year when lambs are being reared. If this is the case, the goats are excluded from the paddock, in order to minimize cross-infection, for goats and sheep have certain diseases and parasites in common. At these times, the goats are tethered on areas of the lawn and on strips of grass bordering the vegetable garden. The tethers are of the swivel-type so that the chances of tangling are minimized. Water buckets in movable metal stands are placed in an accessible spot within their tethered area.

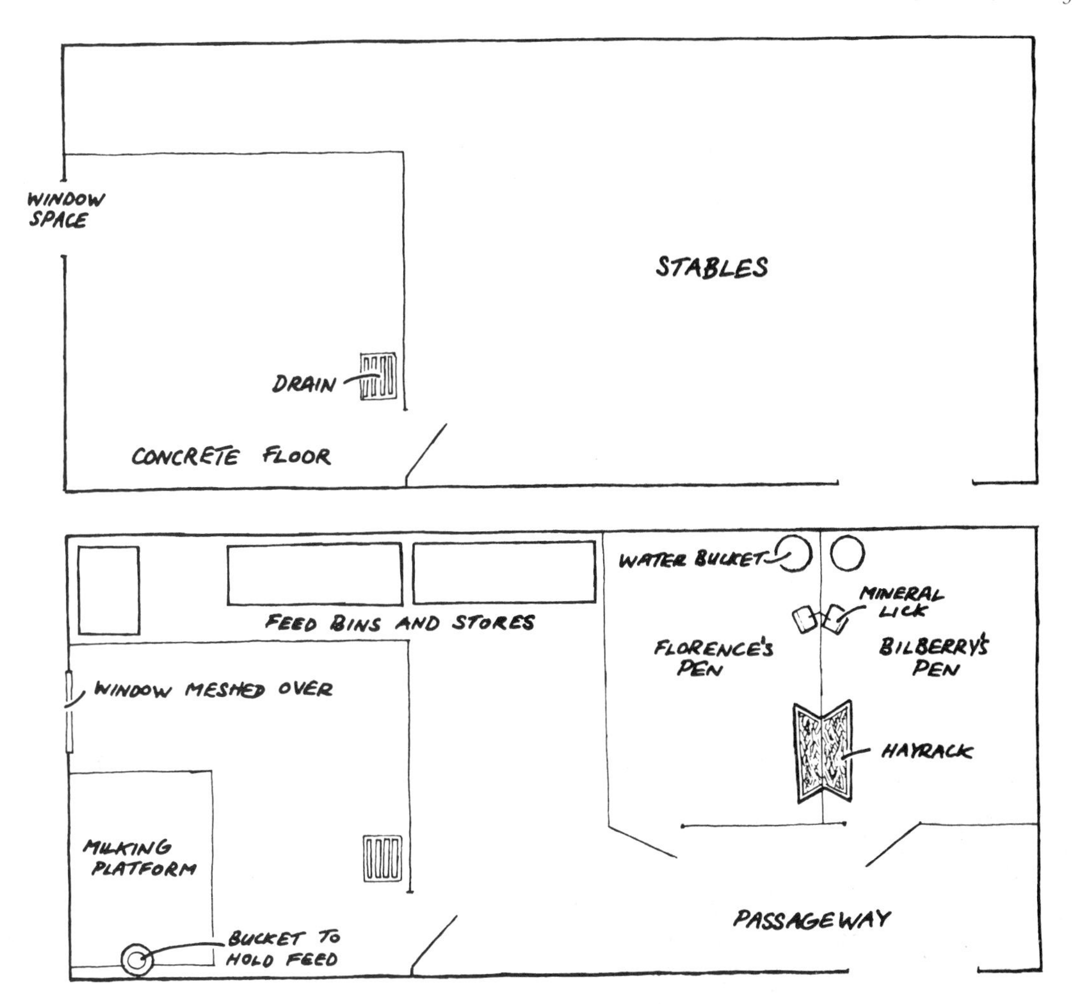

Outbuildings at Broad Leys before and after conversion to goat house

The goat corral

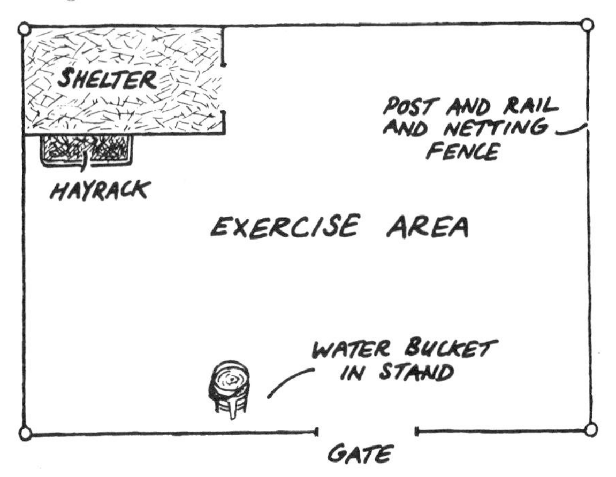

Feeding

Our goats are given half of their concentrate dairy ration in the feed bucket of their milking stand every morning. This is a proprietary feed formulated to give the correct balance of nutrients for a milking goat. After being milked, the goats will either go into the corral or will be tethered on grass. In the corral, they have hedge clippings and other browse material such as elder, willow branches, comfrey and brassica leaves placed in the outside hayrack, and this is added to at various periods during the day. If they are on grass, they are moved periodically so that a fresh patch of grazing is available. In the afternoons, they are brought into the milking parlour for the second milking of the day and are given the remaining half of their concentrate ration. From here, they are taken back to their stalls for the night and hay is placed in the hayracks. Water is available in the stalls, the corral and wherever they are grazing, but not in the milking parlour. A suspended

The goat corral at Broad Leys, with outside rack for browse material and a simple shelter

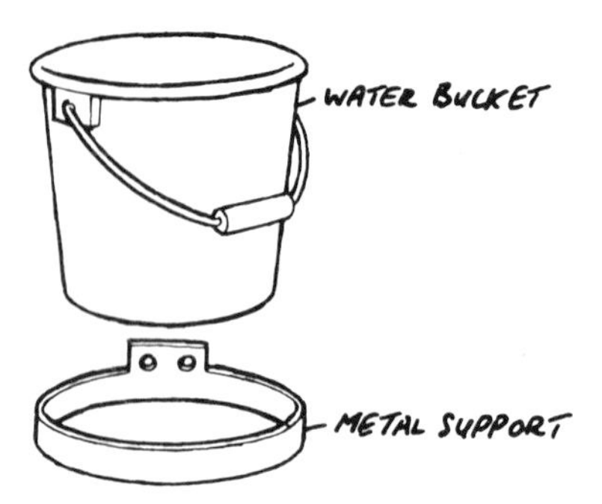

Providing water for inside goats. The water bucket fits into the metal support bolted onto the wall

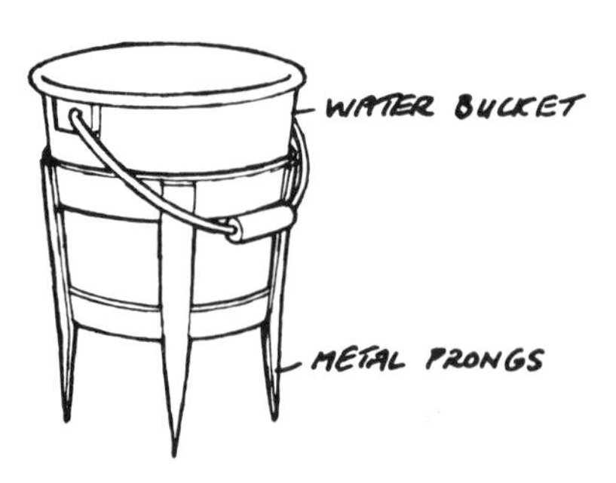

Providing water outside. The metal prongs are pushed into the ground to keep the bucket stable

If there is a weak point in the fence, goats are sure to get through

mineral lick is provided in their stalls so that there is no possibility of mineral deficiency occurring in the diet. In periods of really severe weather the goats do not go out at all, but remain in their stalls. In this situation, they will receive all their rations here, including as much browsing material as it is possible to collect.

Milking and Dairy Practice

Ideally, milking should take place twice a day, morning and evening in clean, dust-free conditions, with the milker wearing an overall. For a small number of dairy animals, handmilking is the usual practice, although milking machines are now available for cows, goats and milk sheep. The larger or commercial units are more likely to have invested in equipment of this kind, although it must be said that there are several commercial goat units which continue to hand milk.

Milking is one of those tasks which is best learnt from someone experienced, rather than from a book (there are now many excellent courses available where such skills are taught).

The udder should first be wiped and the foremilk or first milk squeezed into a 'strip cup' for examination. A 'strip cup' is a special cup manufactured with a black internal surface so that close examination for clots, blood spots and other indications of mastitis and infection is possible. The first few squirts will also contain bacteria and so should be discarded. Once this is done, the rest of the milking can take place as quickly as possible. A stainless steel bucket is the best container to use, not only because it will last a lifetime, but because it is easy to keep clean.

The milk should then be filtered and cooled immediately. Purpose-made filter units are available, but for home use a relatively small quantity of milk can be poured through a kitchen strainer with a paper filter disc in it. Special dairy filter papers are readily available and should be used in preference to other types. The easiest way to cool milk is to stand the metal churn or can containing the filtered milk in a sink and to run cold water over it. For larger quantities, a purpose-made cooler would be more appropriate, and is available from dairy suppliers.

Once cooled, the milk can be bottled, put in waxed paper cartons, or even plastic bags if it is to be frozen for storage. Goats' milk will store for up to three months in a deep freezer, but some commercial goats' milk producers claim that a domestic deep freezer does not achieve low enough temperatures to be effective, and have invested in commercial ones.

Goats' milk should not have taints or 'off-flavours', nor should it be recognizably different from any other milk. If it is there is something wrong with the feeding of the goat, the goat itself, dairy hygiene or storage facilities. If the milk is stored too long, then it will develop taints and the reason why some people believe that all goats' milk tastes 'goaty' is because they are probably buying stored milk which has been defrosted, or so-called fresh milk which is several days old.

We aim to drink milk which has been produced either that morning or the evening before. In periods of hot weather, when it tends to go off more quickly, we pasteurize it. This, on a small scale, is quite simply heating a saucepan of milk to 66 °C (150 °F), pouring it into bottles, standing them in a sink of cold water, then putting them in a refrigerator until required. Ordinary milk bottles are not normally available to the goat-keeper, because they are the property of the local dairy, but lemonade bottles with tops are a good alternative. As heating the goats' milk makes cream rise to the surface, we normally skim this off and put it in a bowl in the refrigerator. After three days there is enough to make butter, or it can be used in any of the ways in which cream from cows' milk is used.

Breeding

A good goat will lactate into a second year, with the yield gradually declining. As the milk from one goat is sufficient for our needs, we have Florence and Bilberry mated on alternate years so that, as one yield is declining, the other takes its place.

It is not worth keeping a billy with a small number of goats. He needs his own quarters and exercise yard, smells abominable and eats a lot. For a specialist breeder the situation is different, and one or two stud males are appropriate.

A goat will come into heat from September onwards and the signs of bleating and sideways tail-wagging are unmistakable. When this happens, our practice is to ring the nearest appropriate breeder with a stud billy and arrange to transport the nanny to it immediately. Transporting a goat is not difficult and a small van or estate car is suitable, as long as there is a dog guard across the back, to stop the animal lungeing forward into the passenger seats. The stud fees for goats are quite

Sunshine, a British Saanen kid aged eight weeks, the daughter of Florence

reasonable and, if the goat does not 'take' the first time, it is normal to have a free service the second time.

The gestation period is five months and births are usually straightforward. If there is any difficulty, however, the vet should be called out. We avoid giving too much concentrate for the first few days after the birth, in case an excess of milk is produced, resulting in calcium deficiency in the mother.

Billy kids have little commercial value and should either be put down at birth or raised for a few months before being slaughtered for meat. The meat is like that of a long, lean lamb and can be used in all the ways in which lamb is cooked. Female kids can be sold locally, preferably through the local goat society, and should be allowed to feed off the mother for five days before being bottle-fed. Dried milk substitute such as that used for orphan lambs is suitable. The first few days of a mother's milk is vital to the young, because of the colostrum it contains. This is milk which is thicker than normal and contains many nutrients and antibodies which give protection against disease.

Goat kids are delightful little creatures and bottle feeding them is a task which children enjoy. Further details of bottle feeding are given in the sheep section (page 106). It has often been remarked that goat kids and children have an affinity and anyone who has ever seen them together would agree with this observation. We have periodic visits from the local nursery school when the small children come to see all the livestock. The goat kids invariably put on a 'performance' for them, leaping, capering and twisting sideways in the air, bringing screams of delighted laughter from all the children.

Regular trimming of goats' feet is essential

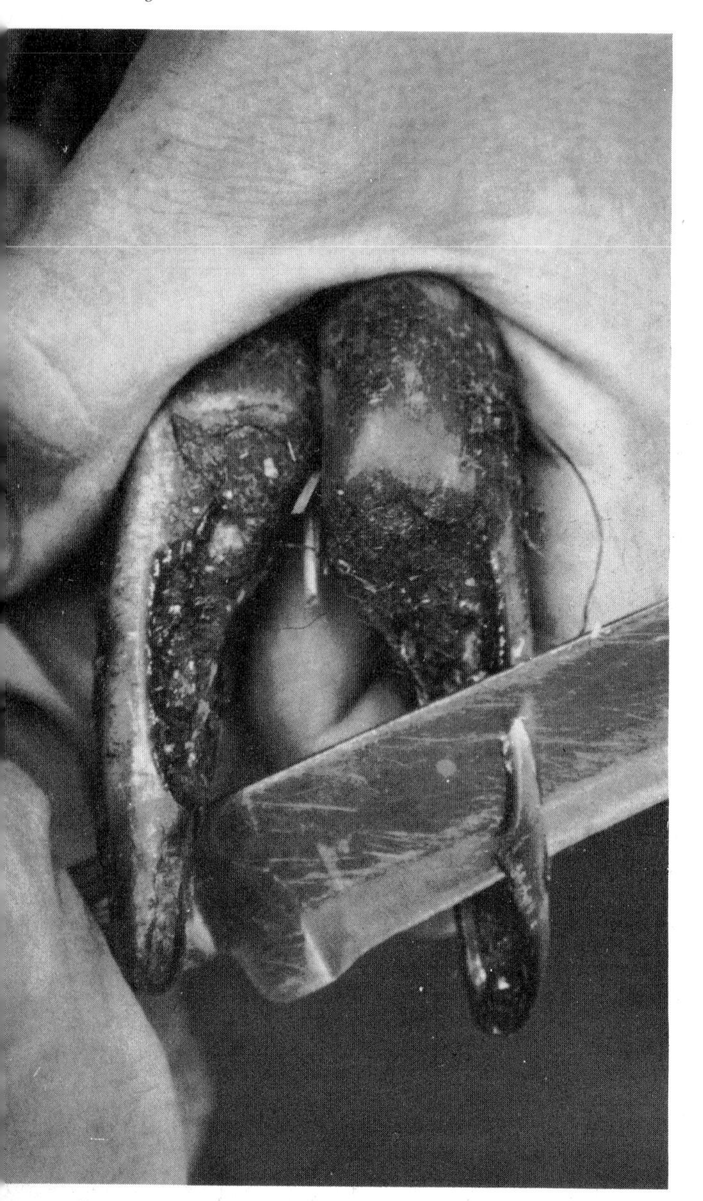

If not trimmed, the nails will curl under the hoof

Health

Goats, like all livestock, need regular worming and this should not be neglected. The most common form in which a vermifuge is administered is as a drench or liquid. With the aid of a helper, the goat's head is held up while the drench is inserted through a tube via the side of the mouth. Injections are also available for this purpose. Vermifuges are needed at least twice a year and possibly four times, depending on the use of the land.

At the beginning of the season, before they go out to grass, all the goats should have an anti-clostridial injection to give them protection against clostridial diseases which also affect sheep.

Kids should be disbudded during the first week: in other words, the horn buds should be burnt out so that horns do not grow. It is best to get a vet to do this. At the same time, contact the local goat-keeping society to find when the next ear-marking session is to be held, so that kids can have identification marks tattooed in the ear.

All goats need their feet checked about once a month, so that overgrown nails can be cut. Once badly overgrown ones are dealt with, a regular filing with a surform blade will keep them in good order.

Heavy milkers are susceptible to the bacterial infection mastitis, causing lumps in the udder and clotting in the milk. It must be treated with antibiotics and the milk discarded for three days after the last treatment. Further details will be found in the section on cows (page 104).

Like all stock, goats are liable to pick up external parasites such as fleas, lice and ticks. These are easily dealt with by the application of veterinary insecticide.

Finally, it must be emphasized that people who keep goats or any of the larger livestock (cattle, sheep or pigs) must keep an up-to-date Movement of Livestock Record Book. This is to record any movements on to or off your premises of animals which may be taken to shows or for servicing. Printed books are available, but an ordinary exercise book will do. What is important is recording the details of the stock's name or number, the date, the place moved to and where from. This is not an unnecessary chore, but of great importance in the control of disease. The local police have the power to ask for your book for inspection at any time, so do not be caught out.

9 The House Cow and Beef Calf

The gentle Jersey – an ideal house cow for the smallholder

For those with enough land, a house cow may be more appropriate than goats. The Friesian is the most common dairy cow in Britain, while the Holstein gives a greater volume than any other breed; but the best cow for the smallholder is undoubtedly the beautiful Jersey. She is smaller than other breeds, with the exception of the very small Dexter, and is placid and friendly. The milk is of good quality, with a high butterfat content, making it particularly suitable for cream, butter and cheese production.

Milk and milk products from a cow cannot be sold without a licence, so there is no question of being able to dispose of a surplus at the gate, in the way one can with goats' milk products. If there is too much of a surplus of milk, it can be used to fatten pigs.

Housing

Cows are extremely hardy creatures and often prefer to be out in severe weather. This does not mean that sleeping quarters should not be made available. They need dry, draught-proof housing with straw to lie on. A simple barn is adequate, but it should be sturdy and well maintained. A good cow provides so much that the least she deserves is a proper place of shelter. Her housing needs are similar to those of goats (see page 90), in that there should be a sleeping area, a milking area and a grazing area.

Feeding

A dairy cow needs good quality grass, for this is the main constituent of her diet. She also needs good hay and a coarse dairy ration, as well as fresh clean water at all times.

A good feeding routine is similar to that detailed for goats (see page 93). The morning milking can coincide with the feeding of half her concentrate dairy ration. After milking, she goes out to pasture and has grass and fresh water during the day. The second half of the dairy ration is given at evening milking time, and when she goes back to her quarters, hay can be provided in a rack. In winter, when grass is not available, hay will be needed to replace it, so obviously the amount consumed in the winter months will be considerably higher than it is in summer.

The amount of feed given is directly linked to the production levels of the cow, although she cannot exceed her own genetic potential. In other words, you can gain maximum production by feeding properly, but once that is achieved, there is no point in feeding extra and hoping for yet more milk. Basically, she needs a maintenance ration to cater for normal metabolism, with a production ration on top, linked with the amount of milk she is producing. In the commercial dairying world, this formula is worked out precisely so that maximum production is achieved. The smallholder who is only catering for family needs will not need to aim for such high production, not only because the feed bills will be high, but also because there may be more milk than is necessary. In addition, mastitis is much more likely to develop under high production conditions. It is therefore up to the individual smallholder to work out how much milk he needs a day and tailor his feeding accordingly.

Use and Care of Pasture

As grass is so important to the dairy cow, it needs to be properly maintained and regarded just like any other crop, not something which magically re-appears each year. Mention has already been made of the fact that there is no feed value in grass in the winter and allowing stock onto it may churn it up and cause damage. During this time, the cow can either be yarded, so that she has sleeping quarters and a yard for exercise, or she can be given a small section of fenced-off field, leaving the remainder of the pasture to rest. The early part of winter is the time to apply lime to pasture, then in late winter or very early spring, a nitrogen fertilizer can be applied to give the grass a boost.

When the new grass begins to grow, take precautions against scouring (diarrhoea) by feeding concentrates or hay before the cow goes out. She should also have a mineral lick suspended in her house, so that minerals are available in her diet. At this time of year, magnesium can be short in grass, and a condition known as 'grass staggers' may result from a deficiency.

The best way to make economic use of grass is by using an electric fence to control access to pasture. Strip grazing, where the fence is moved a short distance further on each day, is probably the best system of arranging this for small acreages.

Milking

The routine for milking cows is practically identical to that detailed for goats, the only difference being that it takes at least twice as long because there are four teats on a cow, and only two on a goat. When handmilking, it does not matter which two teats are milked at a time – whichever combination is comfortable for the milker. Purpose-made milking stools are available and these are made at just the right height for a cow.

Dairy hygiene is paramount, and as soon as milking is completed, the milk should be filtered and cooled. Dealing with greater volumes, as one is with cow's milk, these processes are not as simple as for goats. Specially deep sinks, in the form of plastic troughs on stands, are available, which can hold fairly big churns while cold water runs over them to cool the milk. They are also useful for washing out the bulky equipment such as filter units, churns and milking buckets which ordinary kitchen sinks cannot readily accept.

Individual pens in a small milking parlour suitable for dairy cows

A deep trough sink suitable for the small dairy

The Beef Calf

A cow cannot produce milk for much longer than nine months, and she must calve again in order to start a new lactation. It is not necessary to keep a bull (nor is it wise to try, for they can be extremely dangerous). Fortunately, it is not even necessary to transport the cow to a stud bull, which is just as well, for unlike a goat, she cannot be popped in the back of an estate car. Artificial insemination is widely available and it is merely a question of watching out for the signs of her being on heat – much bellowing and a slight discharge from the vulva. The local Ministry of Agriculture Office (address in Yellow Pages) will be able to arrange for the artificial insemination to take place and the cow should be secured in a stall or barn before the operative arrives.

The Jersey does not produce particularly good beef calves if she is mated to one of her own breed; the flesh has a yellowish tinge which commercially is disliked. For the smallholder it is perfectly satisfactory, but if you want a beef calf with a market value, it is best to choose sperm from a beef breed such as Aberdeen Angus or Charolais. Both male and female calves from this cross will produce good beef animals. If a replacement Jersey heifer is required as a future milker, then sperm from a Jersey bull will be needed. It is a matter of hoping that the progeny turns out to be a female calf. If not, the bull calf is probably best castrated and reared as beef for the home freezer.

The gestation period is nine months and normally there are no problems, but the vet should be called immediately if there is any sign of difficulty. Straining to no avail may be the sign of a breech presentation and professional help is essential. I well remember the telephone ringing late one night, and waking up to be asked by a frantic voice from somewhere in Scotland, 'What shall I do? My cow is calving!' My response, before slamming down the telephone, and couched in polite language for the purposes of this book, was 'Call the vet, you fool!'

After the birth, concentrates should be withheld for the first couple of days in case heavy lactation brings milk fever, a condition where calcium deficiency can lead to collapse and sudden death. If discovered in time, the condition can be reverted by giving a calcium injection, but quick action to call out the vet is essential.

The calf will need to be fed by the mother for about four or five days but after this, should be weaned onto a bucket method of feeding. This is not as difficult as some may suppose. The calf will readily suck someone's fingers and it is simply a question of allowing it to do this while gradually lowering the hand into a bucket of milk

An Angus × Jersey beef calf

until it tastes the milk. Usually, it accepts the idea quite easily, but be prepared for snorts and a shower of milk drops. The calf and the mother have to be separated at this stage, and there is usually a great deal of fuss and bellowing, but after a few days it will cease. Some people prefer to allow the calf access to the mother for part of the day so that they only need to milk once a day instead of twice; they are, in effect, sharing the milk with the calf. Another common practice is to use another, perhaps older cow, as a foster mother, thus releasing the younger cow to continue as a member of the milking herd. If the foster mother is a particularly prolific cow, she may even be able to take several calves; a practice known as multi-suckling. The advantage of this type of rearing over bucket feeding is that the calf is less likely to be affected by digestive problems such as scouring.

We do not keep a cow, for as mentioned earlier, our acreage is too small. We do, however, buy in a calf from time to time, so that it can be raised to provide beef for the freezer. It is an excellent way of using grass to the best advantage. There is a great deal of meat on a beef animal, usually far too much for one family. Our practice, therefore, is to buy a half-share in a calf with another family. We supply the grazing and the day-to-day care, while the others pay for the transport and killing costs. If there are any other costs such as supplementary feeding or veterinary bills, these are borne equally.

If the calf is a bull calf, then it will need to be castrated. Normally, when calves are bought in, they have been castrated already. A sheltered house is essential. This should be rain- and wind-proof and have a good, dry floor with a thick layer of straw for bedding. Any outbuilding will do as long as these conditions can be provided. If an outbuilding such as a barn is too big, the best idea is to pen off a small section in one corner. Sheep hurdles or some sort of similar movable structures are ideal. Straw bales are often placed outside these, to make draught-proof walls. Our practice is to make half of our hay and straw building available to the calf, as we do for lambs when these are raised. This is a stout wooden building with a substantial floor. It faces south and leads out onto the pasture so that, as soon as the calf is hardy enough to go out, access to grass is straightforward.

Where calves are fed on milk substitute, it is important to follow the manufacturer's instructions in relation to dilution. Goat's milk which is an excellent substitute will not need to be diluted unless it is Anglo-Nubian milk. This has a higher butterfat content and will need slight dilution, to the order of 1 part water to 4 parts milk. At first, the young calf will probably take about 1 litre (2 pints) of milk heated to blood heat, three times daily, but this will soon increase to about 4·5 litres (1 gallon). From about a week to ten days, it will also begin to take small amounts of hay. A calf weaner ration can be introduced from about two weeks onwards and as the concentrate and hay ration is increased, the milk ration will gradually decrease. The important thing to remember is to make all changes to the diet gradual so that the chances of digestive upsets are minimized.

Access to fresh, clean water is vital and the easiest way of providing this is to have a bucket in a stand in the house. The type illustrated in the section on goats (page 94) is ideal. A hayrack is also necessary to hold the hay clear of the ground. A suspended mineral lick should also be made available to make sure that the calf has enough minerals in its diet. Concentrates such as the calf weaning ration can be given in a bucket.

Introduction to grass should be gradual in order to minimize the risk of scouring from the lush early grass and it is always wise to feed hay or concentrates before allowing the calf to go outside. Electric fencing is the ideal method of controlling grazing and our system is similar to that detailed in the sheep section. The calf shares the grazing with geese who benefit from having the grass cropped so that new, young growing shoots are made available to them.

It is up to the individual smallholder how long the calf is kept before either being sold or slaughtered. Up to the age of 12–14 months, the resulting meat is called 'baby beef' while that from 20–24-month-old animals is referred to as 'prime beef'. We have ours slaughtered at just over a year because there is insufficient grazing and the cost of increasing amounts of bought-in feeds would make it uneconomical. Where there is plenty of grazing, it is probably worth keeping calves until they reach the 'prime beef' stage, but it should be remembered that over-wintering costs where grass is not available will be comparatively high. Barley straw can be used to supplement the winter diet.

Health

Mention has already been made of the dangers of scouring. Other problems which may arise are external parasites such as lice, mites and ticks – all of which can be dealt with by the use of a veterinary dusting powder. Between September and October, all cattle should be treated with a prescribed skin wash as a precaution against Warble fly, a burrowing pest which is a great nuisance.

Regular worming with a prescribed vermifuge is essential. These can be administered as drenches via the side of the mouth, or in the form of injections. The easiest way of administering such medication is to make a pen out of two gates. Two stout posts are sited about 1 m (3 ft) apart, with the space in between boarded up. Two gates are then placed one on each post, so that when closed they provide a convenient pen which will keep an animal in close confinement. A third stout post will be needed to make closing possible, while a chain barrier keeps the swinging gate secure.

Like most stock, cattle will need to have their feet regularly checked, and the hooves trimmed and rasped.

The most common problem, as with all dairy animals,

is that of mastitis. This is particularly prevalent with heavy yielding stock. It is an infection which affects the udder and the most common symptoms are clots in the milk and lumps in the udder; there may also be bloodstains in the milk. The use of a strip cup to examine the first milk drawn from the udder is an important means of identification at an early stage and should never be neglected. Where it is diagnosed, antibiotic treatment is necessary. This involves milking out the infected quarter of the udder and discarding the milk, then inserting the antibiotic into the teat canal by means of an intramammary syringe. Milk from that area should continue to be discarded while treatment continues and for three days after the last treatment has been given.

Diagram of an udder

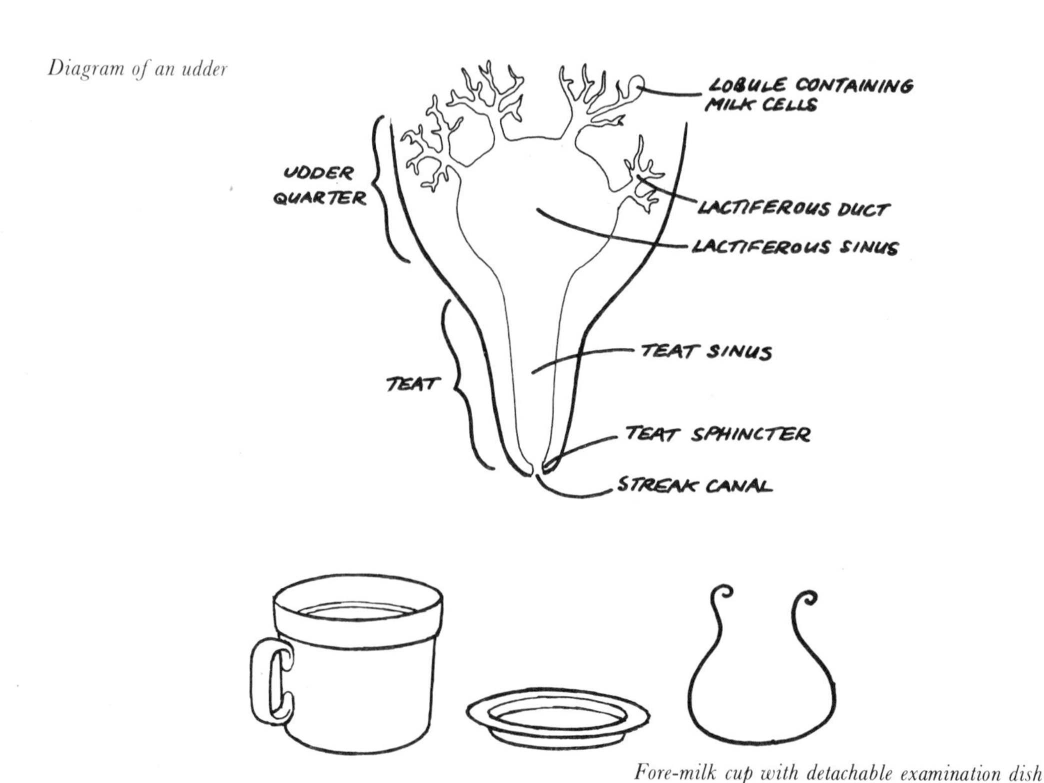

Fore-milk cup with detachable examination dish

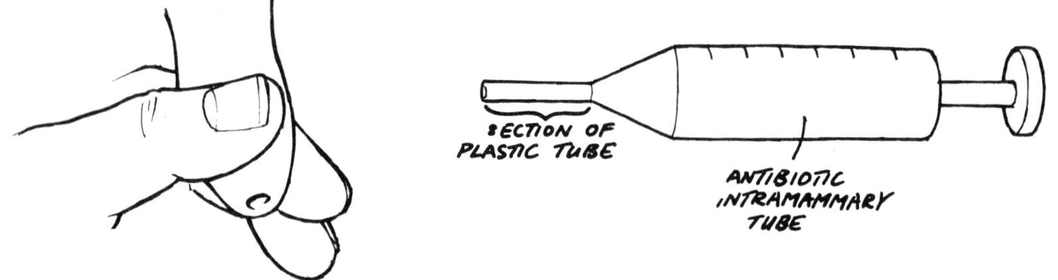

Treating mastitis: administering an antibiotic after milking out. The plastic tube is inserted into the teat opening

10 Sheep and Pigs

Sheep

Different breeds of sheep have been developed for different terrains, with hill breeds such as Shetland and Herdwick adapted to the bleak or mountainous areas of Britain. Lowland breeds such as the Southdown and Romney are heavier and were traditionally found in the milder, more low-lying areas. There is a tremendous variety in the breeds available and there has been much cross-breeding over the years. The choice of breed for the individual depends primarily on the nature and amount of land, the availability of stock in a particular area and the reason for keeping them. Most smallholders will probably be keeping a few sheep to control the grass and to provide meat for the home freezer, skins for curing, or wool for spinning. Any breed of sheep will cater for all these necessities, of course, and it is an individual choice. What it is important to remember is that while the more primitive mountain breeds such as Shetlands will adapt to milder, lowland conditions, the opposite is not always true, and some lowland sheep may find it difficult to adapt to bleaker, mountain conditions. The type of sheep will also determine the kind of fencing used. The more modern cross strains have shorter legs than the older types and can be confined with 90 cm (3 ft) sheep netting. More primitive and long-legged breeds such as the Jacob, which are popular with home spinners, will need at least 1·5 m (5 ft) because of their tendency to jump. Sheep will also try to push through a barrier so there must be no weak points in the fence. The main food of sheep is grass so there must be adequate supplies.

Sheep on a Small Scale

Our decision to keep sheep was based on the fact that we had an acre of grass which was not being adequately eaten down by grazing geese and browsing goats. Geese concentrate on the growing tips of fine young grass shoots while goats prefer a range of fine grasses and broad-leaved weeds. We needed to have stock which would also eat the more coarse grasses, otherwise the pasture would decline over a period of time. It would not do for us to have sheep permanently, but it was appropriate to raise a couple of lambs from time to time, and a beef calf more occasionally.

We had tried taking hay from our acre, but as it would have taken too long to have cut, turned, and carted it by hand, we had a contractor to do it. By the time he had cut, turned and baled it for us, each bale cost us precisely what it would have cost to buy in. We decided to buy hay in future and to use the grass to fatten lamb or beef.

With a small area of land it is not worth keeping breeding stock, or indeed any sheep that are being kept from one year to the next, because of the dipping regulations. These involve immersing, and keeping immersed for one minute, all sheep which are not being slaughtered that winter. The dipping tank needs to be of Ministry-acceptable design and depth, and a prescribed strength of dipping solution must be used. These regulations are necessary to control the incidence of sheep scab, a condition which can have a devastating effect on the sheep's wool. It is too much of an expense for the smallholder with just a few sheep to buy and instal dipping equipment. However, it is possible to share the facilities of a larger sheep farmer who may either charge a small fee or come to some other mutually acceptable agreement. Any such movement must, of course, be recorded in your Movement of Livestock Record Book.

If sufficient land is available, and a small breeding unit is being established, the best time of year to acquire stock is in late summer or early autumn. For the smallholder with limited land, it is more appropriate to acquire two or three lambs in the spring, possibly orphan ones which are sometimes more cheaply available, and to raise them for the freezer. If obtained in late spring, they will not need complex buildings, and a simple shelter to give protection against wind and rain is all that is necessary. Our lambs are not given special housing, but a small area of the stable in the field where the hay and straw supplies are stored, is used to house them. Access to the straw and hay bales is blocked, and they are able to go out of the stable into the field at will.

The grazing is not made available all at once, for we find it better to rotate it. In other words, an electric fence is used to partition the field into several small plots. The lambs go onto the first plot and are then moved on when the grass is eaten down. After a few days, the geese go onto the first plot because they like short grass and new young shoots are revealed by the close cropping of the sheep. This procedure is continued for the rest of the grass growing season. A water trough, regularly replenished, should also be made available.

A Southdown ewe with a newly born lamb

Feeding

When lambs first arrive, they will need to be bottle fed on a dried milk substitute, made up according to the individual manufacturer's specifications, for there are differences between the various trade products. Alternatively, goat's milk can be given, without dilution; if fresh cow's milk is given, it should be diluted to the ratio of 1 part water to 3 parts milk. The milk should be at blood heat and a few drops poured on the inside of the wrist will indicate whether it is too hot or too cold. Rubber lamb teats are available from suppliers, either as plug-in types or pull-on ones. Both will fit onto lemonade bottles and the teats are suitable for goat kids as well as lambs.

The age of the lambs will indicate how many feeds they should be getting. The smaller they are, the less they will drink at a time, but the more frequently they will need to be fed. They will probably start with four feeds a day, of about half a bottle at a time, but you can quickly gauge their appetites and cater accordingly. Any milk left over should be discarded or given to the cat, and not warmed up again for the next feed. The number of feeds will gradually be reduced to three, while giving more in the bottle, and then to twice a day, when each lamb will probably drink a whole bottle at a time. By this time they will also be taking solid food. Introduce them to a few lamb pellets or some coarse ground sheep mix, both of which contain added vitamins and minerals. Then, gradually allow them onto an area of clean grazing. They will soon begin to nibble. There is a danger of scouring (diarrhoea) from eating grass which is too lush, as it often is in the spring. The best way to avoid this is to make some pellets or 'mix' available to them before they go out and to restrict their access to grass to short periods for the first few weeks. A little hay can also be given before they go out on grass. If scouring does occur, keep the lamb penned in a sheltered place on straw and give no milk, only clean water in a small trough. A little concentrate ration can also be given. Ordinary scouring from lush grass will soon clear up, but if the cause is a bacterial infection (possibly from not sterilizing the feed bottle properly) then an antibiotic injection from the vet may be necessary.

The period of bottle feeding will last until the lambs are about two months old. It is time-consuming, although many would say enjoyable, but children are fond of helping out with this particular task. It should be made quite clear to them, however, that the lambs are not pets, but livestock destined for slaughter when they are heavy enough. This may seem hard, but it is the reality of the situation. Children are far less sentimental and more realistic than is often realized. They appreciate truth and will respect it. It must be admitted that sentimentality in relation to killing livestock tends to come from urban adults, rather than from country people.

Meat Lambs

As long as there is sufficient pasture and fresh water for the growing lambs, their needs will be few as they continue growing into the summer. The main thing to watch out for is blowfly attack or 'strike' as it is often referred to. This is where a blowfly lays its eggs in a small cut or where dung has adhered on wool at the back. Maggots quickly hatch out and will penetrate and feed on the sheep's flesh. Watch for signs of an animal stamping, or nodding its head vigorously, and take quick action. The section of wool around the affected area should be clipped back and the maggots extracted with tweezers. Clean out the wound with soap and water to which household disinfectant has been added, then apply antiseptic ointment to prevent infection and to discourage further attack.

When the grass begins to decline in late summer, the lambs can be slaughtered for the freezer. If the weather is still hot, it is best to send them to a local slaughterhouse for butchering. If they are to be killed at home, it is best to wait until the weather cools.

It is legal to slaughter livestock on your premises if it is done humanely and instantaneously and there is no distress caused to the animal during the period while it is awaiting slaughter, or during the killing process. The meat must be for private consumption only and must not be sold. Only that which has come from an officially-licensed and inspected slaughterhouse can be sold.

I would not advise smallholders to kill their own larger stock. It is much better to get an experienced and licensed slaughterer to do it for you. This need not involve having to transport the stock to the abattoir. Local self-sufficiency groups are increasingly organizing a system where a 'freelance' butcher will visit a smallholding to slaughter stock. Our own area has such a system and I believe this to be preferable to using a slaughterhouse where a certain amount of distress is inevitable, for the animals can smell the blood when they arrive.

I can vouch for the fact that home-slaughtered animals die quickly, painlessly and contentedly in familiar surroundings. Our practice is to give the animal a little food to eat and while it is concentrating on this and enjoying it, the slaughterer quickly places a humane killer on its forehead, between the eyes, and the released bolt renders it totally unconscious instantaneously. As soon as this happens, the throat is cut and the carcass is suspended head-downwards to bleed. Death really is instantaneous and there is no suffering.

After skinning and disembowelling the carcass, you should hang it in a cool room for about a week before it is jointed and placed in the deep freeze. The method of dealing with the skin is described on page 123.

The Breeding Flock

As mentioned earlier, a breeding flock is rarely possible only for those with sufficient pasture to support such an enterprise. Ewes and a ram for breeding should be bought in autumn and properly checked over to see that they are in good condition. Check the feet to make sure they are clean, trimmed and free of infection; the mouth in case there are any broken teeth, and also check the age. There should be no discharge from the eyes, nostrils or mouth and the rear should be free of dung. They should have been dipped in the approved manner that autumn and the vendor should provide a certificate to show this. Check that they have also been wormed and are up-to-date with their anti-clostridial injections.

When they arrive on your land, run them through a footbath containing either a 6 per cent formalin solution or a 10 per cent copper sulphate solution. A footbath can be constructed as a permanent fixture by making a shallow, concreted depression in the ground, or a movable, galvanized bath can be purchased from livestock suppliers. Whichever is used, it will be necessary to have sheep hurdles leading to it, so that the sheep can be controlled and directed into it. As sheep do not like getting their feet wet, they have to be persuaded; this is where a sheep dog comes in useful. The object of

Local self-sufficiency groups often organize a freelance butcher service for the humane home slaughter of livestock

the footbath is to prevent footrot, and ideally, the sheep should be run through once a month.

The new ewes are 'flushed' or brought to peak condition for breeding by making sure they have enough to eat, including any late pasture that has not been grazed. If they have not already been dipped, this should be carried out. Every year, the Ministry of Agriculture announces two dates, indicating the period during which compulsory dipping must be carried out. It must be in an approved manner, for a specified time and using an approved dipping chemical. An official will check that the procedure has been carried out correctly and will then issue a certificate accordingly.

Soon after dipping the ram can be introduced to the ewes and mating will take place. With a small number of sheep it will probably be possible to observe that all of the ewes have been served. Where larger numbers are involved, it is a good idea to fit the ram with a sire harness which has the effect of marking the backs of the ewes with coloured dye when they are mounted. By changing the dye colour regularly, it will not only be possible to see which ewes have been serviced, but also those which have been serviced to no avail, if they are marked with more than one colour. If all the ewes are multi-coloured, change your ram quickly for he is obviously infertile.

Once all the ewes are in-lamb, they will need to be given over-wintering rations of hay to replace the grass. A concentrate ration can be given to build up their reserves and they should be brought to a sheltered place such as a barn before lambing. Sheep hurdles or straw bales can be used to partition it off into individual lambing pens. Give them all another injection against clostridial diseases, so that the unborn lambs also receive protection, and worm them a few weeks before lambing or 'tupping' takes place.

Once the lambs are born, they should have their tails docked. This is a matter of placing a tight rubber ring around the tail, leaving just enough section to cover the vent. The ring cuts off the blood circulation and after a few weeks, the tail drops off. It may seem a cruel practice, but if not done, a great deal of dung can get caught on the tail, making it a breeding ground for maggots. Ram lambs are castrated in a similar way by using an elastrator or piece of equipment which stretches and places a rubber ring around the scrotum. Any lambs which are to become breeding stock rather than meat lambs should be earmarked for identification.

Shearing

June is the traditional time for shearing sheep and it is not difficult to do this. What is difficult is to do it well. For a small number of sheep it is better to do it yourself, but for larger numbers, it is more appropriate to have it done by contract shearers. If you have five or more sheep, you are not allowed to sell the wool other than through the Wool Marketing Board.

After shearing, the sheep are dipped again; this time as a precaution against external parasites such as lice and ticks.

Pigs

I cannot pretend to be fond of pigs, although I would be the first to admit how intelligent they are by comparison with other livestock. I have a wild, childhood memory of a large, miry monster that looked as if it would have eaten anything and, in fact, did once consume an unwary chicken that had flapped its silly way into the sty. Such associations have a lot to answer for, because I still find it difficult to overcome a deep distrust of pigs.

Of all livestock, pigs are probably the ones most likely to draw complaints from neighbours, unless one is meticulous in clearing up the muck. Pig manure has a particularly objectionable and penetrating smell, although in fairness it must be said that it is usually the large scale, intensive unit which is most to blame in this respect. Large units have a considerable problem with muck clearance and usually end up having to store it in slurry tanks which are periodically emptied. In small quantities, the droppings can be cleared away at regular intervals and stacked with straw in order to rot down. This provides excellent fertilizer for the kitchen garden when incorporated during winter digging.

Pigs are large, strong animals which can be difficult to confine. Housing and enclosures need to be strong in order to stand up to them. I had always resisted the idea of keeping them until we came across the Pot-Bellied Vietnamese pig. This breed is about a third of the size of a normal pig, black in colour and completely 'undeveloped' from the selective breeding point of view. It is therefore a 'primitive' pig, but exceptionally hardy. One of the drawbacks of the modern, commercial hybrid pig is that, although it produces a large amount of good quality, lean meat, it is also much more liable to succumb to infection than its more rugged ancestors.

We decided to keep this breed partly as an experiment, because a small pig seemed to be less fearsome and easier to handle. Its size was also more appropriate to the needs of a family and the feeding costs would be far less than for a bigger pig. So we acquired two pigs, a male and a female, from two different wildlife parks. It was necessary to record the movement of these in our Movement of Livestock Record Book, but in addition, pigs need a permit from the local office of the Ministry of Agriculture before they can be moved. It is quite straightforward to arrange this and the permit normally comes through in a matter of days. The reason for the added regulation is to try to cut down the risk of disease.

Other breeds which are popular with smallholders are Gloucester Old Spot, Tamworth and Saddleback. The Gloucester Old Spot is a hardy pig which, in the past, was often kept in orchards in order to benefit from the

Pigs as ground cultivators, confined with electric fencing in one section of the vegetable garden

windfalls. The modern hybrid is based on the Landrace, a long, lean animal which has been bred to produce less fat and more lean meat.

Housing and Management

In the corner of our kitchen garden there is a brick building about 2 × 6 m (6 ft 6 in. × 19 ft 8 in.). It is called the pump house because there is a small well inside which used to supply water for Broad Leys and several neighbouring houses. The pump and machinery had gone when we came, so it was a matter of sealing the well shaft and clearing the debris from the floor. A good, sound concrete floor was revealed and the building, with its door, two windows and corrugated metal roof, was soon ready for occupation.

Around the house we constructed an enclosure of wooden stakes and pig netting. The latter had to be really well anchored into the ground to prevent the pigs lifting it up with their noses. The fencing does not need to be particularly high, for pigs tend to push their way through or under an obstacle, rather than over it.

Feeding

Feeders and drinkers are important because they need to be shallow for access and heavy for stability. Pigs are experts at overturning or even destroying lightweight containers. As we happened to have an old earthenware sink, we decided to use this as a drinker, for it is easy to keep clean as well as being extremely heavy. Pigs, like all livestock, need to have fresh, clean water all the time.

We bought a new, heavy metal feeding trough which is designed for shallow access and stability. We had first considered feeding the pigs on proprietary pig pellets, but soon realized that the expense was too high, so we compromised by buying in bags of 'sow and weaner meal'. This is mixed with water to a porridge-like consistency and additions such as vegetables can be incorporated. It is important to avoid all kitchen scraps which incorporate meat because of the dangers of disease. Regulations prohibit the feeding of swill to pigs, unless the swill has been treated in licensed premises.

Home-grown fodder crops such as turnips and the outer leaves of vegetables such as brassicas are valuable additions and help to reduce the overall feeding costs. Fruit is also well received, as long as it is fed in moderation; too much at a time may cause digestive upsets. The Vietnamese pigs are the only livestock we

Vietnamese Pot-bellied pigs

have ever encountered that will eat banana skins. Where potatoes are fed, they should be cooked first before being added to the meal. Where there is a surplus of home-produced milk, skimmed milk, or buttermilk as a result of buttermaking, it can be used instead of water to mix with meal. Piglets grow at a phenomenal rate when given this diet, with its inexpensive protein content.

Pigs can be kept on grass and there are many people who prefer to do this. We decided against this because we preferred to keep the grass purely for grazing stock. Where it does take place, a house will need to be provided. There are several good, portable structures on the market which are suitable. These houses can also be used when pigs are confined to a section of the kitchen garden so that their ploughing activities can be used to bring about winter digging. Temporary fencing should be rigged up to confine them to a small area at a time, and then they can be moved on to the next area as soon as the first one is ploughed up. Portable electric fencing is the easiest method of arranging this.

Breeding

Many smallholders find that breeding pigs is too specialized a subject because it involves keeping a boar which can be dangerous to the inexperienced handler. It is much simpler, if the aim is purely to produce pork for one's own freezer, to buy in a couple of piglets from a breeder and raise those. This is normally our practice and the only reason we had a boar was for the purpose of experimenting with the Vietnamese breed. As it happened, our particular boar was as friendly and gentle as could be wished, and he did much to dispel my dark attitudes in relation to pigs in general.

The gestation period of pigs is about 115 days, and normally, the gilt (or unmated female) is allowed access to the boar only when mating is desired. As far as our two pigs were concerned, they were inseparable and resisted all efforts to house them separately, so we allowed them to share the same accommodation until a few days before the piglets were due to be born. At this time, we made a temporary farrowing (birth) house in another outhouse. This required only a clean, sheltered situation, straw on the floor as bedding and a rail sited in such a way as to divide the room into two; a large section for the sow and a smaller section for the piglets to which the sow would not have access. The piglets, being smaller, can go under the rail to suckle the sow as necessary and then return to their own area. An infra-red lamp hung over the piglets' area ensures that they return there and that they are kept warm. The reasoning behind this system is that it provides protection for the small piglets in case the sow rolls on top of them and smothers them.

The female was quite ready to move to her new quarters and walked as sedately as she could, despite the fact that her belly was, by this time, touching the ground. The boar made every effort to come as well and was obviously disturbed at being left on his own. We had to make a point of going to see him many times during the day and scratching his head behind the ears. This was popular and brought many grunts of pleasure.

The birth took place without any difficulty. It is always best to be present while a sow is farrowing so that, as each piglet arrives, it can be checked over to make sure its mouth and nose are clear of mucus. It is a good idea to place it temporarily in a cardboard box with a piece of old blanket placed over a hot water bottle. This keeps it warm until its brothers and sisters join it. It also ensures that it is not harmed while the sow is giving birth to the next piglet. Ten piglets were born, with one dead at birth. Another which was smaller and more puny than the others died two days later. Each was checked over and given a brisk rub with an old towel before being put in the cardboard box.

Once the sow had cleared up the afterbirth, we placed the piglets near her teats and they immediately began to suckle. After a meal they soon discovered the warmth of the infra-red lamp and went under the rail to curl up in a small, black heap. They were about 15 cm (6 in.) long at birth and a beautiful, glossy black. As they grew, they gradually became more of a charcoal grey colour, like the parents.

The sow proved to be an excellent mother, making and remaking her nest with immense care, carrying straw in her mouth to place in the appropriate position. She was also careful to leave dung only in one area of the room, from where it was easy to remove it every day for adding to the compost heap. In commercial units, sows are placed in metal farrowing crates where they are kept immobile. Even after the birth has taken place they are still confined in this way, unable to turn, while the piglets suckle and then go back to their own quarters. Many people find this system distasteful because of the lack of humane consideration, and the small-scale pig-keeper certainly has no need to follow this practice. In large units, there is little personal relationship between the sow and the handler, while on a small scale, the pigs soon learn to trust their keeper. Our sow showed no signs of disturbance when we picked up one of her babies in order to stroke it; in fact she would come and nuzzle us with her nose until we stroked her too. It is worth mentioning that she succeeded in getting her nose under the rail after two days and lifted it up from its supports. We found her, one morning, lying under the infra-red lamp with all the piglets fast asleep around her. She showed immense care when settling herself down and there was obviously no risk of her lying on and smothering the piglets. We removed the rail and never used it again. It should be emphasized however, that a rail is recommended, for individual sows vary in their behaviour, and it is best to see how she behaves before deciding whether to discard or retain it for their safety.

Piglets feed off the mother for the first few weeks, but will quickly show an interest in the sow's food, as well as the drinking water. The proprietary 'sow and weaner' meal is excellent in this respect because it means that one does not have to bother with the preparation of separate rations. As the piglets grow, they consume more solids and less milk until they are naturally weaned at about 7–8 weeks. In large units it is the custom to give iron injections to piglets during the first week of life in order to counteract anaemia. The tails are also docked so that they do not injure each other by chewing them, while the sharp canine teeth are clipped so that damage is not caused to the sow's teats. It should be remembered that in intensive units, abnormal behaviour patterns develop because of the unnatural and overcrowded conditions in which the piglets are kept. We have never had occasion to dock the tails of piglets, because they were always kept in humane conditions, with plenty of room and enough to interest them. We have never had a case of chewed tails either. The decision to clip the teeth is an individual one. We have never done it because there was never any sign of soreness on the sow's teats, but, in cases of uncertainty, it is best to obtain veterinary advice. The same applies to the question of iron injections. They are needed by the piglets in large units because they do not go outside and are thus unable to obtain minerals from plants or from the soil. Our practice, like that of many small pig-keepers, is to place a section of turf in the piglets' quarters during the period they are confined. They will chew and root through this very happily. Once the weather is warm enough, they can go out into an enclosure or into a field with the sow. Here they usually have access to a range of plants and earth to cater for their mineral needs. Vegetables mixed into the feed will also ensure that their diet is balanced.

We sold our piglets once weaned, keeping two to raise for the freezer. Pigs are normally raised to pork or to bacon weight. In a normal large breed, the former is about 45–70 kg (100–150 lb), while the latter is about 70–90 kg (150–200 lb). In our case, the weights achieved were smaller because of the small nature of the breed, but it is worth adding that when we took the decision to end our experiment with the Vietnamese breed, the boar was 45 kg (100 lb), while the sow was 40 kg (88 lb). Most smallholders tend to raise their weaners to pork weight because of the extra feeding costs involved in bringing them up to bacon weight. If it is done, it is better to have the meat cured by professionals rather than try to produce bacon oneself. While it is legal to slaughter a pig on one's own premises if the meat is for home consumption only, it is essential to obtain the

Young pigs can either be sold as weaners or raised as porkers

services of an expert. The animal must first be stunned with a humane killer before the throat is cut. A licensed slaughterer will ensure that it is carried out expertly and humanely. Further details of this are given in the section on sheep.

Coping with Problems

The main emphasis, as with all livestock, is on prevention, and a lot of complaints can be avoided by keeping a careful watch for problems. Mention has already been made of the need to ensure that the diet is balanced, without an excess of one type of food, and that there is access to fresh water. Any signs of scratching or soreness on the skin is probably an indication of mite attack. A dusting with an appropriate veterinary powder, or a wash with a liquid preparation, will remedy the problem.

Breeding stock will need to have their claws clipped back from time to time, and as they are strong, professional aid may be needed, unless the pig-keeper is skilled and experienced in the technique. Purpose-made clippers and rasps are available from equipment suppliers.

Pigs which are out on grass or in enclosures will need

Moving a pig – even a small one – can be difficult

regular worming and the vet will advise on different types of vermifuges.

One problem which may be encountered is how to catch a pig. This is obviously not a veterinary question, but it is one that crops up frequently. I have already said that fencing needs to be strong in order to confine a pig. Even so, there is always the occasion when you suddenly find that a pig is in a place it should not be. The only way to herd a pig along is to obtain the help of friends, each one armed with a wide piece of wood or galvanized sheet of metal; a dustbin lid is also quite effective. A pig will dive straight at a person, endeavouring to escape through his legs, but if a wide obstacle is presented, it will deter it.

Often, it will be unnecessary to have to resort to the above, for food is a great incentive and a bucket of food will sometimes succeed when all else fails. On one occasion, we had a gilt who was on heat and escaped from her pen. We found her one morning, standing stock-still, like a statue, outside the kitchen door. She could not be persuaded by any of the normal inducements, and even a smack on the rump had no effect. This is a normal pattern of behaviour for the female will remain like this until mounted and mated by a boar. What the gilt failed to appreciate on this occasion was that we did not have a boar. The only way we were able to move her was to lift her back legs and place them a few inches further away, do the same with her front legs, then continue in this way until she was back in her enclosure. The whole operation took nearly two hours and is not recommended for anyone with a tendency towards impatience. While this behaviour pattern was a great nuisance to us at the time, it is most useful for anyone who does not have a boar, but wishes to utilize artificial insemination. The local office of the Ministry of Agriculture or the vet will be able to advise on the nearest A.I. centre which is able to supply frozen sperm. Insertion with a catheter is quite straightforward.

11 The Smallholding Harvest

Harvest is traditionally recognized and celebrated after the cereal crops have been safely gathered in. For the smallholder it is not always a time of plenty: it is sometimes a matter of luck with the weather and the incidence of predators. It may depend on whether you remembered to close the gate of the kitchen garden to keep the chickens out, or how late you were with your planting. There are always years when some crops do well, while others are a disaster, but there is usually something of surplus proportions that is worth preserving.

Some crops, such as potatoes, apples and carrots are stored 'fresh', while more perishable foods have to be subjected to some form of preservation technique in order to make them last. There is now a great deal known about food technology and preservation methods: information that our forbears did not have. Some of the older methods are now known to be hazardous and we decided, early on in our smallholding activities, to forgo the production of home-cured bacon and preserved smoked foods, preferring to leave these to the professionals. In fact, when it came to preserving our harvest at Broad Leys, we initially looked at and tried all the different methods so that comparative evaluation was possible.

Freezing

This is now the most common method of preserving food at home, and justifiably so, for it is a quick and clean method which does not rely on additives or other preserving agents. The food is pure and reconstitutes with little deterioration in quality. The initial capital outlay in the purchase of a deep freezer is high, but prices have fallen in recent years as they have become more popular. A chest freezer is more economical than an upright one where there is more temperature fluctuation when it is opened. A freezer should be opened as infrequently as possible so that these fluctuations are kept to a minimum; it is not too difficult to plan ahead and decide what needs to be taken out, and when.

The ideal place for housing a freezer is in a cold outhouse or garage so that there is a saving in the amount of energy required to reach the appropriate temperature. Where this is not available, a thick blanket or similar material will provide insulation. It is also worth remembering that a full freezer is more economic of power than a half empty one. A good way of filling in the gaps, on a temporary basis, is to put loaves of bread in the freezer.

Mention has been made earlier in the book of the greater tendency to have power cuts in the country than in urban areas. When this happens, the freezer should not be opened at all, but kept insulated with thick blankets. The contents will suffer no deterioration for many hours, normally allowing sufficient time for the fault to be corrected. Most of the worst cuts tend to be in winter when the outside temperature is low anyway, but, for those worried at the prospect of losing their stores, there are now special insurance schemes which make provision for this unlikely event.

There are many accessories available from freezer suppliers, but, while many of these may reduce the time spent on some activities such as food preparation, they are not essential. In our experience, plastic bags and ties are the only really essential items, although plastic baskets which stack and slide along in the freezer itself make storage and retrieval of frozen packs much easier.

Unless vegetables are being used in a short period, within a few weeks, they will need to be blanched. This is a process of subjecting the surface to a high temperature for a short period so that bacteria and moulds are killed off. The easiest way of achieving this is to immerse the prepared vegetables in boiling water or steam for a few minutes, before cooling and freezing them. If this is done, there is no danger of taints developing in the food.

There are several ways of blanching and cooling. Many people use a blanching basket, rather like a chip pan, to dip the vegetables into boiling water, and then run cold water through it. This is perfectly satisfactory, but does tend to make the finished product slightly soggy. We prefer to blanch in steam. The vegetables are placed in a stainless steel colander which is then placed on top of a large saucepan of boiling water. The lid is placed on the colander and the steam then blanches the food for the required time, depending on the size and density of the individual vegetables. For relatively soft vegetables such as Brussels sprouts, a blanching time of around three minutes would be required if they were immersed in boiling water. If steam blanching was used, they would need half as long again, a total of four-and-a-half minutes. Carrots need about five minutes in water and seven-and-a-half minutes in steam. Particularly soft vegetables, such as courgettes and leeks need about two minutes in water and three minutes in steam.

Once the blanching is complete, the colander is removed from the heat and left on the draining board for a few minutes while a plastic bag is placed in a previously scrubbed earthenware plant pot. The blanched vegetables are then placed in the plastic bag, conveniently supported by the plant pot, and then quickly transferred to a bowl of ice-cold water. The water on the outside of the bag has the effect of forcing out the air and a tie is quickly placed around the neck of the bag. Once the bag is cool, it is labelled and placed in the fast-freezing section of the freezer. All produce should be dated.

Soft fruit, such as gooseberries and blackcurrants, which are likely to be used fairly soon, do not need blanching and are merely placed on trays which are then placed in the freezer. Once the fruit is frozen, the tray is removed and the fruit placed in individual storage bags.

Meat merely needs jointing before it is bagged and frozen, while poultry and rabbits can be frozen whole. It is important to defrost completely before cooking, unless a micro-wave oven is used. The danger is that the outside is cooked while the inside remains uncooked and Salmonella poisoning may result. This is more usually associated with whole poultry than with other meats.

Salting

Salting was a common preservation method before tinned foods became widely available. After that it declined and freezing has made it even more uncommon. There are people who still do it, but it is now comparatively rare. Rock salt is required because the larger crystals ensure dry packing and there is less of a tendency for the product to go off. Runner beans were the vegetables most frequently preserved in this way and anyone over the age of 40 probably remembers the crockery jars that stood on pantry shelves, including the ones with beans that became so slimy that they had to be thrown away.

Rock salt is now expensive to buy, and is no longer the cheap, easily available commodity it was in our parents' generation. I have preserved beans by salting several times in the past, but now no longer do so because it is cheaper and less time-consuming to freeze them.

The beans need to be quite dry and placed in jars, with a thick layer of rock salt separating them from the next layer. They can either be stored whole, with just the ends trimmed, or chopped into sections, depending on the size of the container. Crockery jars are now no longer readily available, but large glass storage jars will suffice, as long as they are stored in a dark pantry. Beans will store for several months in this way, but must be well soaked to remove the salt before they are cooked.

Drying

If the moisture content of food is removed, decay is prevented, but there has to be a desiccating or drying agent to bring this about. In sunny areas of the world there is an abundance of free solar energy, but, in Britain the sun cannot be relied upon to make an appearance when needed. Purpose-made dryers are available, but normally the running costs of these make the operation non-cost effective, unless large quantities are being dried. Beekeepers, who use small, home-made solar wax collectors for recycling their wax supplies, have found them to be equally effective in drying small quantities of fruit such as apricots or apple rings.

In Britain, on a domestic scale, the usual practice is to utilize the surplus heat rising from an Aga or Rayburn solid fuel cooker. Drying trays made of muslin stretched over light wooden frames are suspended in such a way as to take advantage of the rising convection currents of warm air. Apple rings are often slipped onto pieces of bamboo canes or dowel rods and either dried in this way or placed in a cool oven. Dipping them in lemon juice, vinegar or brine stops browning while this is taking place, and also keeps flies at bay. The main problem is in ensuring that, once dried, the food does not then reabsorb water so that moulds appear. Most dried foods sold in the shops have preservatives to stop this happening and, traditionally, a sulphur dioxide candle or cone was used; the fumes would impregnate the food and provide protection. Anyone with an interest in wholesome foods and in restricting the amount of additives in the diet, would find this a dismaying practice.

Drying herbs is a different matter, for they are thinner and will dry more quickly than denser products, and as long as they are kept in dry storage containers with a tight-fitting lid, will not normally become damp. They can be placed in a cool oven or put above a range, but I find that an equally effective method is to hang them in the conservatory where they gain the maximum amount of sun and are not affected by rain. Those that are grown for their seeds, such as Caraway, have paper bags placed over the seed heads to catch any falling seeds, but it is important to remember to pierce air holes in the bags, to provide a flow of air.

Making Jam

Making jam is not difficult, but it requires attention to detail to produce a good quality product that retains its colour, and is neither too runny nor so solid that a spoon can hardly penetrate into it. The basic principle of jam making is in first cooking the fruit slowly in a quantity of water, so that by the time it is soft, the natural pectin or setting agent has been released. Where natural pectin is limited, as it is in fruits such as strawberries, a high pectin agent such as lemon juice is added. Commercial pectin is also available, but should not be necessary for home purposes.

Once the fruit is cooked, the required amount of warmed sugar is stirred in and the mixture heated as *quickly as possible* to the setting temperature of 105 °C (220 °F). Stirring should take place all the time to prevent the jam sticking to the bottom of the pan. As soon as the setting temperature is reached, the jam is removed from the heat, allowed to cool for a short time, and then put into warmed jars. Clean jars placed on their sides in a cool oven, with the door open, will be sufficiently warmed and sterilized without the danger of cracking. Once they are ready, they are placed on a wooden board while the jam is ladled in; a purpose-made funnel simplifies this. The jars are then sealed, labelled and stored. This, then, is the basic procedure, with minor adaptations depending on individual recipes; but there are several things that can go wrong.

The most common mistake is to try to make too great a quantity at a time. Where this happens, the jam bubbles away for ages, without ever reaching the setting temperature. If the setting point is not reached, the jam will be runny in the jars and is likely to go mouldy. It is often not realized that the traditional jam pan, which is still sold in many shops, is not suitable for some modern stoves. On some electric cookers, for example, insufficient heat is generated to cope with a large amount of jam in the open pan; too much heat is lost from the surface before the overall temperature is high enough. It is much better to make smaller quantities in an ordinary stainless steel or aluminium saucepan such as would normally be used for potatoes. In this way, the setting temperature is reached in a short time, enabling one, for example, to produce gooseberry jam that is green, not a pale, insipid pink because it has been boiled too long.

Another point at which things can go wrong is in not knowing the precise moment at which the setting temperature is reached. The traditional advice is to put a spoonful of jam on a saucer and then tip it sideways. If the surface of the jam wrinkles, then it is ready. While this is true, it is still a rather hit-and-miss technique, and is really only appropriate for those with long experience. In my view, it is much more sensible to purchase a jam thermometer, rather than rely on eighteenth-century hearsay. Jam thermometers are readily available and provide a foolproof source of information.

Sometimes, despite everything, a film of mould will develop on the surface of stored jam. It is not harmful and once scraped off leaves the jam quite edible, but it is unsightly. A good tip which I had from my mother (and she from hers) is to dip the wax sealing disc in vinegar before putting it on the surface of the jam. It does not affect the flavour and it really works.

Ideally, preserving sugar should be used for making jam. Granulated sugar tends to form a scum on the surface, but this can easily be skimmed off and it does not have an adverse effect. I always use granulated sugar because it is cheaper and more readily available than the coarser preserving type, but I make jam for family use, not for show.

In areas where the water is hard, calcium salts may have the effect of making the jam cloudy, but, again, this does not affect the quality of the product, merely its appearance.

There are many recipes for making a wide range of jams, so I will confine myself to giving details only of those which are particularly popular in our family.

Rhubarb and Ginger Jam

1 kg (2 lb) rhubarb
1 kg (2 lb) sugar
2 lemons
12–25 g (½–1 oz) root ginger (according to taste)

Make this jam with the later, end of season rhubarb rather than with the earlier sticks which can be bottled or made into pies and crumbles. Wash, trim and chop the sticks and place in a bowl with each layer covered by sugar. Add the juice from the lemons, then leave the bowl to stand overnight. This is one of the exceptions to the normal way of making jam where the sugar is usually added after the fruit has been cooked.

The following day, put the rhubarb, lemon juice and sugary syrup in a saucepan and add the root ginger. Bring rapidly to the boil and cook until the setting temperature of 105 °C (220 °F) is indicated on the thermometer. Remove the ginger, if preferred, although this is not strictly necessary, and pot into jars. Seal and label with the date.

Gooseberry Jam

1·5 kg (3 lb) gooseberries
2 kg (4 lb) sugar
600 ml (1 pint) water

Choose firm, green and slightly under-ripe fruit; that from a cooking variety is better than that from a dessert variety. Top and tail the gooseberries and put in a pan with the water. Stew gently until the fruit is soft before adding the sugar. Recipes always call for warmed sugar, without ever telling one how to achieve this, or without taking human forgetfulness into consideration. It is easy to forget until the time comes to add it to the fruit. My practice is to try and remember to put the sugar in the airing cupboard overnight. With most jams, it is not crucial to have warmed sugar, but with gooseberry jam, the aim is to boil for as short a time as possible so that the green colour is retained.

Once the sugar is added, boil the jam as fast as possible until the setting point is indicated on the thermometer. Remove from the heat immediately this is achieved, then pot, seal and label.

Pickles and Chutneys

Pickling is another well-known preservation method, where the presence of vinegar inhibits the growth of bacteria. Pickled onions are favourites in our house and the shallot onions, ready in late July and August, are grown especially for this purpose.

The best thing about pickled onions is eating them, but the worst part is undoubtedly peeling them in the first place. I have never been able to find an easy way of doing this. At the best of times it is a tearful event, and the quicker they can be peeled the better. Once peeled they should be placed in a bowl and covered with brine, made up in the proportion of 50 g (2 oz) salt to each 600 ml (1 pint) water. Leave for two days, topping up with extra brine if necessary, and be sure to keep the bowl covered with a clean cloth to prevent flies and dust falling in.

After two days, wash the onions really well in cold, running water, otherwise they will be too salty to the taste. Meanwhile prepare the vinegar which can be white or malt pickling vinegar. Place in a stainless steel or aluminium saucepan, with 25 g (1 oz) pickling spice per 600 ml (1 pint) vinegar. If, like us, you like your pickled onions fiercely spiced, then increase the amount to 50 g (2 oz).

Bring the vinegar to the boil and keep at boiling point for five minutes. Allow to cool, then pour over the onions which have been packed into jars. As we like our pickled onions hard, we start to eat them after a week, but many people prefer them soft and so leave them for a few months before sampling. The onions will gradually soften and become browner in storage.

Chutneys are made from a mixture of fruit and vegetables, vinegar and spices and differ from pickles in that all the ingredients are cooked together. The best approach is to use whatever happens to be available in glut proportions, but as it is difficult to generalize, I shall give details of just one which is a favourite every year.

Red Tomato Chutney

1·3 kg (3 lb) tomatoes
450 g (1 lb) onions
450 g (1 lb) cooking apples
225 g (½ lb) soft dark brown sugar
10 ml (1 heaped or 2 level tsps) salt
600 ml (1 pint) malt vinegar
50 g (2 oz) pickling spice

Peel and chop the onions and put in a saucepan and cook with half the vinegar. Meanwhile, put boiling water into a dish and drop in the tomatoes, one by one, so that they are easy to peel once removed, and put in cold water. Chop the peeled tomatoes, then peel, core and chop the apples. Add the tomatoes and apples to the onions and pour in the remaining vinegar, together with the brown sugar and salt. Place the pickling spice in a piece of muslin tied into a bag so that it can be removed after cooking. Bring to the boil, stirring all the time, and cook until the chutney thickens. Cool and put into jars, seal and label. It is delicious with cheese and cold meats.

Bottling

The principle of bottling is to subject the produce to a sufficiently high temperature to sterilize it and then seal the container so that air is excluded. It can only be used for fruit because the acidic content is high enough to prevent potentially dangerous food poisoning, but it is not safe to try to bottle vegetables at home; the risks of botulism are too great. The exception is the tomato which does have sufficient acid and can be safely bottled.

Bottling jars with purpose-made sealing lids are readily available, as well as replacement lids and sealing rings. There is a choice of jars, but one which I have found to be satisfactory has a graduated scale on one side, with lids that are plastic coated to prevent a chemical reaction and rubber seals. A threaded ring seals the lid tightly on the jar.

The so-called water bath method is probably the most common way of bottling and is certainly the method which I find most satisfactory. A deep saucepan, fish kettle or boiler is needed to hold the jars. It needs to be wide enough to hold several jars, side by side at the same time, and deep enough to ensure that the jars are completely submerged. An inner grid is required to keep the bottoms of the jars from direct contact with the pan base. The only other equipment needed is a thermometer which will register the boiling point of water, 100 °C (212 °F), and bottling tongs which are made in such a way as to grip the necks of the jars, making removal from the hot water an easy task.

Fruit can be bottled in water or syrup, but there is a tendency for fruit to rise in the jar when syrup is used, although the colour is better. Syrup is made by dissolving 225 g (8 oz) granulated sugar in 600 ml (1 pint) of water and bringing to the boil, then allowing to cool.

The fruit is washed, peeled if necessary, and cut into appropriate sized pieces. It is packed into the jars, using the handle of a wooden spoon to push the pieces down, but taking care to avoid bruising. Add the cold syrup or water right to the top of the jar and bang it gently to release any air bubbles. Put on the lids and rings, with the screwband in position, then unscrew the screwband by a half turn to allow any remaining air to escape during the heating process. Place the jars in cold water, making sure they are submerged, then bring to a temperature of 74 °C (165 °F) in about 90 minutes. Keep the jars at this temperature for a few minutes, depending on the type of fruit. Soft fruit will need ten minutes, while large, hard fruits require 20–30 minutes. This method is referred to as the 'slow water method', and requires a thermometer to ensure accurate temperatures. A quicker method, which does not need a thermometer, is the 'fast water method'. This is where the fruit in the jars is topped up with hot water or syrup, and hot water is placed in the pan. This is brought to simmering temperature, 88 °C (190 °F) in about 25–30 minutes and held at simmering point for the required time, two minutes for soft fruit and sliced hard fruit, ten minutes for stone fruits and 30 minutes for hard or large whole fruits.

Once the bottling process is complete, remove the jars with the tongs and place on cloth or folded newspaper. Tighten the screwband and leave for 24 hours. The next day, unscrew the band and remove it. Lift up the jar by the lid; if the vacuum is complete, it will hold. Any jars which have cracked or chipped rims will be unable to sustain a vacuum and should be discarded. The fruit in jars which have not been successful should be used straightaway. Jars for storage should be labelled and stored in a cool, dark pantry.

Storing Fresh Fruit and Vegetables

Crops such as potatoes, carrots, apples and pears can be stored fresh in cool protected conditions, so that they are available in the winter months when the taste of summer has all but gone. The important thing to bear in mind is that only the later 'keeping' varieties will store. There is no point, for example, in trying to make an early eating apple last: the result will be a crabby, dried-up object in a very short time. It is also important to provide suitable conditions for storage. The ideal is a cool, airy situation, free of damp, and where the temperature does not fluctuate, but stays at a few degrees above freezing point. The traditional root cellars provided these conditions, as did the old apple stores which were insulated with straw and had thatched roofs of the same material. For most people, a garage or outhouse is a suitable alternative, but in recent, severe winters, these have often been inadequate, and it is a good idea to make provision for this by having insulating materials such as straw, old newspapers or pieces of old carpet to give extra protection.

Mice can be a great pest where crops are stored and every effort should be made to deter them, including, if necessary, the placing of traps. Where the store is outside, slugs and snails may find the clamp to be an over-wintering haven, with food laid on for them.

Potatoes and Carrots

Root vegetables were traditionally kept in an outside clamp. This involves placing a thick layer of straw on the ground and then putting the crops on the straw. A further layer of straw covers them, followed by soil which is tamped down on top, although a few wisps of straw are left projecting at the surface in order to provide 'chimney' aeration. This system can work well as long as the winter is not too severe, and provided that the field mouse population does not learn of the clamp's presence. In our area of East Anglia, we have extremely severe frosts, and our experience is that clamping is too risky a business. We prefer to store our potatoes in paper sacks in a cold outhouse. Wooden pallets are placed on the concrete floor, with straw on top. The sacks are then placed on top of this, with added straw between each sack and on the surface.

Carrots can also be stored in this way, or placed in sand in a bin or wooden box. We freeze a proportion of our early carrots and store the later varieties in paper sacks. Maincrop carrots can also be left in the ground until needed, in the same way as parsnips, but they are at risk from mice and slugs, and will need to have a layer of straw placed over the tops in case of severe frost.

Onions

Mention has already been made of the shallot onions which provide our pickled onions. Large onions, grown from sets, are stored, tied in strings, and suspended in the same outhouse as the potatoes and carrots. These last well into spring and even early summer, although by then they are beginning to get a bit soft. Traditionally, there was always a gap before the first of the new season's crop became available, after the last of the old had sprouted and become inedible. The new Japanese varieties have now filled the gap, with the seeds being sown in April for harvesting in early summer. There is now also a new British variety, Unwins Autumn Planting, which we have found to be even more reliable. This is planted as sets and grows well, even in the most severe of winters.

Plaiting onion strings is not difficult, but it does take practice. There are several ways of doing it and it is largely a matter of individual preference as to which method is adopted. The easiest way is to take a piece of stout string and merely loop and tie it around each onion, adding a new onion every few centimetres. Tie a loop at the top, when it is complete, and suspend the string from a hook. My particular method is the traditional one of plaiting three onion strings together into a rope, and then adding a new onion each time one of the strings becomes short. It is more difficult to do, but, once mastered, is a quick and attractive method.

Only onions which are firm and with a long thin string should be plaited. Any which are 'bull necked', or having a thick area at the top, should be used as soon as possible, as well as those which are in some way damaged. Any which begin to sprout should be removed from the string and used. If the onion itself is too soft to use, the sprouted shoots can be used as a replacement for chives or salad onions.

Apples and Pears

Apples and pears will store well if given the right conditions, but again, it should be emphasized that it is only the hard, keeping varieties which will store successfully. One of the best methods of keeping them is to use purpose-made apple trays in cardboard boxes, but an alternative is to wrap each apple in newspaper before placing it in the box. The advantage of this is that if one apple becomes mouldy, the paper stops it spreading to other apples. Again, a cool, frost-free outhouse is the ideal storage place.

Any fruit which is damaged or bruised should be discarded, for it may have an adverse effect on the rest.

It can still be used for making wine, so that it need not necessarily go to waste.

Making Wine

Making wine used to be a 'hit-and-miss' affair, but, since the booming interest in the subject in recent years, the availability of efficient equipment and sterilizing techniques have made it much easier and more sure of success. Kits are also readily available and it is recommended that the beginner should try these out first before venturing into making wine from his own produce.

Essentially, wine is produced by the fermentation brought about by living yeast cells in a sugar solution. Although yeast, like all living organisms, takes in oxygen and gives out carbon dioxide, it has the added ability of carrying out respiration in the absence of air. This is called anaerobic respiration, where sugar is broken down to form alcohol and carbon dioxide, and is the basis of all alcohol production.

In Britain, there are no legal restrictions on the making of wines and beers, as long as they are for home consumption only. It is against the law to try to produce spirits.

There is no need to spend a fortune on equipment, and as previously mentioned, the purchase of a kit is an economic way of acquiring many of the basic needs. A number of household items can also be used. A large stainless steel or aluminium saucepan or boiler will be needed for initial cooking of the fruit, as well as a plastic bin for the initial storage of the fruit during its first 'mashing' stage. This should be a high-density, non-toxic bin from a wine equipment supplier, rather than an ordinary bin which may leach toxic residues into the wine. Glass fermentation jars with airlocks are really essential, as well as purpose-made wine bottles which are strong enough to withstand pressure from carbon dioxide. Corks or plastic caps will be required, as well as a siphon tube for racking off or siphoning the wine into bottles. Although natural yeasts will provide fermentation in country wines where fruit is used, it is much less haphazard to use a sachet of special wine yeast from a supplier. Other, miscellaneous items of equipment are jugs, wooden spoons, a nylon or muslin strainer and a thermometer. Finally, and perhaps most important of all, Campden tablets are a means of ensuring efficient sterilization of equipment so that the wine does not go off.

Blackberry Wine

The list of possible home-made wines is, of course, endless, but this blackberry wine is a favourite in our house.

2 kg (4 lb) blackberries
1·5 kg (3 lb) granulated sugar
4·5 litres (1 gallon) water
Wine yeast
2 Campden tablets
25 ml (1 level tsp) citric acid

Wash all the utensils carefully and sterilize in a solution of Campden tablets. To produce this, crush the tablets and dissolve in 600 ml (1 pint) of warm water. When cool, add the citric acid.

Remove any sepals still attached to the blackberries, wash them thoroughly. Place in the plastic bin and crush them with the back of a wooden spoon. Now heat up 4·5 litres (1 gallon) of water and, when boiling, pour it over the fruit. Stir well, then put on a tightly-fitting lid and leave to steep for three days. Strain off the liquid through a strainer and add the sugar, stirring well to make sure that it all dissolves. Add the wine yeast which has been previously started with warm water and sugar or yeast nutrient according to the manufacturer's instructions. Siphon the liquid into glass fermentation jars and fit fermentation airlocks. These will allow carbon dioxide to escape, but will not allow the entry of air. Place the jars in a warm place: an airing cupboard is ideal. Fairly fiery fermentation will proceed for a few days, depending on the temperature, but once this has slowed down, siphon the wine into clean jars and top up with a sugar solution made up of 75 g (3 oz) sugar in 600 ml (1 pint) of water. Refit the airlocks and leave the jars in a warm place for a further two months. After this time, siphon the wine into sterilized bottles and cork firmly. Leave for a few months to settle and clarify, then drink.

Finally, it is worth mentioning that one of the great enemies of the winemaker is the small insect *Drosophilla meleangaster*, or vinegar fly. This will introduce bacteria into the wine, causing it to turn vinegary, if it gains access. One of the best ways of avoiding trouble is to ensure that no fruit is left lying around where winemaking activities take place, otherwise flies will be attracted to the place in great numbers.

Smoking

Smoking is one of the oldest preservation methods known to man. In the light of modern knowledge, it is also one of the most hazardous to health, with a considerable risk of introducing carcinogens into the body. This is particularly true with prolonged exposure to smoke, and the traditional practice of 'cold' smoking which was used for farm hams should be approached with great caution by the amateur.

There are two distinct methods: 'hot' smoking which is a short exposure to smoke while the food is being cooked at the same time, and 'cold' smoking where previously brined foods are exposed to cold smoke over a longer period of time. It is this latter method which can be dangerous and the advice, which is still occasionally given, to hang a ham in the fireplace of a farm kitchen is extremely irresponsible. Hot smoking, using one of the many appliances available, is quite safe, as long as the manufacturer's instructions are followed, and we use it often to smoke chicken, trout and even cheese. It does not preserve food for a long period as cold smoking does, but we prefer to forgo this technique because of the potential dangers. If we want to smoke our own ham or bacon, we take it along to a professional curer who does it scientifically and accurately.

Hot smokers are readily available and are made of heavy gauge metal which does not warp with heat. They are normally in three parts – a heat box, baffle plate and food grid – and use sawdust or thin shavings. It is important to ensure that this comes only from well seasoned hardwoods, not from freshly cut or resinous softwoods. Buying it from a smoking supplier will ensure that there is no risk in this respect. Methylated spirit is frequently used as a heat source for hot smokers and does work efficiently and cleanly, but it is important to remember not to tip the appliance while it is in use in case the fuel spills.

Home Dairying

With your own milk supply, you are in the fortunate position of having a source of home-produced cream, butter, yoghurt and cheese. Goat's milk does not produce as much cream as that of a cow, but there is usually enough for family use.

Cream

Cow's milk is left to settle for a day or two so that the cream can be skimmed off the surface. Goat's milk needs to be heated slightly to make the cream rise, then left for a further few hours to cool and settle. The use of a centrifugal cream separator means that virtually all the cream can be removed, but for normal family use it is not worth the considerable expense. The cream can be used in liquid form or will thicken if whipped. Many people do not believe that goat's milk cream can be whipped, but I have frequently served whipped goat's cream to visitors and they have been unable to distinguish it from any other cream.

Butter

Butter is merely cream which has been beaten to the extent that it 'breaks' into butter particles and buttermilk. Most cooks, at some time or other, have whipped their cream too hard only to find to their dismay that the cream destined for the trifle has turned into butter.

An ordinary food mixer can be used to produce butter, but only relatively small quantities are possible at a time because of the splashing. Where butter is made regularly for family use, it is worth investing in a small butter churn, for these are readily available from dairy suppliers.

The cream is left to ripen for about two days and then placed in the churn which should not be more than two-thirds full. It is then merely a matter of turning the handle if it is a hand churn, or switching on if it is electrically powered. Warming the cream slightly beforehand ensures that butter is produced in a relatively short time. The easiest way of doing this is to stand the glass churn of cream in the airing cupboard for a short time. If it is really cold, I pour in a small amount of hot water to increase the temperature. It does not affect the resulting butter and merely dilutes the buttermilk slightly.

After churning, the butter particles form, and the buttermilk is then strained off into another container. It should not be thrown away for it is useful for making scones, as well as providing a refreshing drink if kept for several days so that natural souring takes place.

The butter needs to be washed with cold, running water so that all traces of the buttermilk are washed away. After that, it is 'worked' or kneaded and compressed with a wooden spoon or a pair of traditional butter workers which now once more available from suppliers. The object of 'working' is to remove the last vestiges of water so that a pat of butter can be shaped. If salted butter is preferred, the salt can be incorporated during the 'working' stage.

Goat's milk butter is quite white and often surprises visitors who think that they are being offered lard to put on their bread. Once they have tasted it however, they are forced to admit that there is no difference between it and ordinary butter, apart from the appearance.

Yoghurt

Yoghurt is so easy to make and so nutritionally beneficial that I am surprised that so many people buy it in the supermarket, instead of producing it for themselves. It is simply a matter of heating milk, either cow's or goat's, to a temperature of 82 °C (180 °F) in order to pasteurize it, then cooling to 43 °C (110 °F). This is then poured into a large thermos flask which has previously been rinsed out with boiling water. A dessertspoonful of shop-bought, plain yoghurt is blended with a little warmed milk and added to the thermos. Alternatively, a little commercial lactic starter can be used. This gives more consistently good results, and is available from dairy suppliers. Give the thermos flask a shake to distribute the starter, then leave undisturbed for the yoghurt to form. Once it is ready, it should be placed in the refrigerator to cool and firm. It is delicious on its own, mixed with honey or with fruit.

Soft Cheese

The easiest soft cheese to make is the one which merely involves using up naturally soured milk. This is poured into a colander lined with muslin, and the muslin is then tied up to form a bag which is suspended above a sink or bucket in order to drip. Once most of the liquid has gone, open the muslin and scrape the curd from the sides into the middle and vice versa so that draining is thorough. Suspend again for a few hours, then remove the curd from the cloth and add salt to taste. It can be made more creamy by beating in a little butter and can be flavoured with black pepper, garlic or chives. It is important not to use shop-bought milk for souring in this way, because this has already been pasteurized, and the souring in this case runs the risk of harbouring potentially harmful bacteria. Home-produced milk which has not been pasteurized is satisfactory. Rather than wait for the milk to sour naturally, the process can be speeded up by adding a little buttermilk, vinegar or lemon juice.

Hard Cheese

Making a hard cheese is a much more complicated process and is not economical for anyone who has to buy in milk from another source. With your own dairy animal however, it is an excellent way of utilizing a milk surplus, particularly in the spring when a cow has just calved or a goat is newly kidded.

If hard cheese making is to be a fairly regular event, then it is worth investing in some proper equipment, rather than trying to make do with something else. For example, it is worth buying proper cheese moulds and a press because one of the most difficult aspects of making a hard cheese is in applying sufficient and regular pressure. A large saucepan, boiler or vat is needed for heating up the milk, as well as a dairy thermometer, large knife to cut the curds, muslin to hold them, and miscellaneous utensils such as spoons, jugs and trays. Apart from milk, cheese rennet will be required for curdling, as well as a 'starter' for introducing the culture which gives the cheese its particular aroma. These are available in dairy suppliers' shops. Cheese rennet is essential, but a perfectly good cheese can be made by using a home-made starter. My practice, if I am making a particular cheese, is to take a small piece of shop-bought cheese and mix it up with a little boiled and cooled water, strain it and then use that as a starter. It is best to use a piece of cheese from just below the rind.

Cheddar-type cheese There are many hundreds of different hard cheeses, but one of the most straightforward is the Cheddar. It can be made from cow's milk or goat's milk and there is no truth in the claim that goat's milk will only produce soft cheese. I make Cheddar cheese regularly from the milk of my goats and it is indistinguishable from any other Cheddar (unless it could be said that it is better than most). Here is the recipe that I use.

14·5 litres (3 gallons) milk
125 ml (5 fl oz) starter
25 ml (1 tsp) cheese rennet
25 g (1 oz) salt

Put the milk in the boiler or vat and heat to 68 °C (155 °F). Cool as quickly as possible to 32 °C (90 °F) by standing the pan in a sink of running water. Add the starter and stir it in well. As previously mentioned, this can be a commercial lactic culture or a home-produced one from a piece of shop-bought Cheddar. Leave for half an hour then add the rennet after first diluting it with four teaspoonfuls of boiled and cooled water. Stir it in well and also incorporate the cream at the surface. Leave for between 20 and 30 minutes, by which time the curd will have formed. A good way of testing this is to put the back of the finger on the surface; if it comes away without leaving a white stain on the skin, it is ready.

The curd must now be cut and this is a matter of making vertical cuts, first one way and then the other so that squares are formed. After that, diagonal cuts are made so that the curd is separated into portions. Leave for ten minutes until the whey appears on the surface, then increase the heat gradually so that it goes up to 38 °C (100 °F) in about half an hour. During this time, gently stir the curds and whey by hand. Once the maximum temperature has been reached do not let it exceed this. Remove from the heat and let the contents settle for a further half hour.

Now ladle off as much of the whey as possible and tip the curds into a sterilized (boiled) piece of muslin. Make this into a bundle and place on a tray tilted to one side so that draining continues. After a quarter of an hour remove the curd from the cloth and cut into four long slices. Replace the cloth and leave to drain for a further 15 minutes. Open the cloth again and restack the slices so that the inner ones are on the outside and vice versa. Repeat this process several times until the curd looks rather like cooked white chicken when it is broken.

Break the curd up into pieces about the size of a nutmeg and sprinkle on the salt. While the salt is being incorporated, prepare a new section of muslin by sterilizing it in boiling water, then lining a mould with it. This can be either high density, non-toxic polythene or stainless steel. Some cheese presses come complete with moulds. Press the curd into the mould, then fold over the ends of the muslin. Put the wooden follower of the press onto the top and place in the press with a fairly light pressure. After a couple of hours, increase the pressure a little and then leave the whole thing overnight.

The following day take the cheese out of the mould and replace upside down, using a fresh piece of muslin in the mould. Increase the pressure again. The next day, take the cheese out again and examine it carefully. If there are no cracks in it, dip it for half a minute in hot water at 66 °C (150 °F). Return it to the press again, remembering to reverse its position, and increase the pressure slightly. Leave it in the press for a further five days, turning it at least once a day and increasing the pressure slightly each time.

After five days, remove the cheese from the press and leave uncovered in a cool, fly-proof room, so that it forms a dry rind. After a few hours, it will be ready for binding. Cut a piece of muslin which is long enough to wrap around one and a half times the circumference of the cheese, together with end pieces or caps for the ends. Using lard or flour paste, stick the caps on the ends and wrap the long piece around it firmly, gluing it with the lard or paste.

The cheese is now ready for storage at a temperature of 10–16 °C (50–60 °F), either on a shelf or suspended in a muslin bag. If it is on a shelf remember to turn it daily. It will be ready to eat after about a month, but for a more mature cheese, leave longer.

Curing Skins

One aspect of smallholding life which is likely to crop up from time to time is what to do with the skins of slaughtered livestock such as rabbits, sheep or goats. If the larger livestock have been slaughtered at a registered abattoir, the skins will have been kept, for they are regarded as one of the 'perks' of the job. There is, of course, quite a demand for sheep and goat skins for they make excellent rugs. Rabbit skins can be used to make mittens and gloves or even waistcoats and coat linings by the dedicated home crafts person.

Once skins are removed from the animals, they need to have as much fat and tissue removed from the inner surface as possible. This is made easier by pinning the skin out on a board – an old door salvaged from the scrapyard is ideal. At this stage it will be virtually impossible to remove the slippery inner membrane; it is much easier once the skin has dried out a little. The problem is that while this drying is taking place, the skin can also go off. There are a number of chemicals which can be used at this stage, but for those who do not wish to have to buy them and who want a quick and easy method of working the skins, the old wartime practice of using paraffin is as good as any. Soak paraffin onto the skin which should be lying horizontally on the board, and leave it to dry out. Add fresh paraffin every few days, and, as drying proceeds, begin to scrape away the inner membrane. There are purpose-made scrapers available, but virtually any blade that is comfortable to use is suitable. Pumice stones and sanding blocks have also been used to good effect. Sheep and goat skins can be worked with a circular sanding disc of an electric drill, but care should be taken to avoid snarling the wool or hair at the edges of the skin. This method is unsuitable for rabbit skins. Once all the inner membrane is removed, the skins can be washed well and left to dry. Sheep and goat skins can be washed in the washing machine, and one skin will fit into the spin dryer at a time, but rabbit skins should be washed by hand. When they are completely dry, they will be rather stiff but if rubbed vigorously will soften. More paraffin can be rubbed into the under surface at this stage, and finally a little oil to finish it off.

REFERENCE SECTION

This is a reference section to enable those who are in a hurry to make quick checks.

The Smallholding Year

Many tasks are seasonal and often the problem is in interlocking the jobs associated with different activities so that everything can be fitted in at the appropriate time of year. It should not be supposed, however, that timing is absolutely critical. There is a considerable leeway in, for example, sowing times and these will be affected by a number of factors including weather, geographical location and needs of specific plant varieties. Similarly, fences do not *have* to be repaired in the winter; it is merely that this time of year when animals are frequently inside, is more convenient.

The following tables act as a general guide, for each individual will have his or her own variations and pattern of working, but they enable the reader to have a quick and easy reference guide to help in forward planning. A particularly important aspect is the emphasis on regular disease prevention.

The year is seen as beginning in the autumn, for this is the time of preparation for the coming year, not just a time for clearing up before the winter.

The tables appear by courtesy of *Practical Self Sufficiency* magazine which first published them as a booklet called *The Smallholder's Year*.

The Land

Autumn
September–November

Allow grazing to continue in early autumn, as long as grass is not over-grazed.
Allow stock to graze green fodder crops such as field-sown turnips. Ideally, control with electric fending.
As the weather worsens feed fodder crops inside.
Plan ahead for next season.
Allocate winter quarters for stock and plan next year's grazing and hay areas.
Turn hens onto harvested arable fields to glean in early autumn, but put them in winter quarters by late autumn.
Turn breeding sows onto land that needs clearing so that it is ploughed and manured.
Turn ducks into vegetable garden to clear slugs, but protect Brussels sprouts.
Put geese in orchard to benefit from windfalls (breeding stock, not Christmas geese).
Plough land which is to have a new grass ley.
Harrow, sow and roll the new field.
Cull non-productive stock to avoid expensive winter feeding.
Clear rubbish to discourage rats.

Winter
December–February

Check land drains and clean out ditches.
Repair fences and cut hedges.
Repair, clean out and disinfect outdoor housing which is unused over the winter.
Check and oil tools which will not be used until spring.
Plough arable land if a new ley is to be sown in the spring.
Plough other areas for spring crops.
Apply organic manures.
Apply lime if needed, but not at the same time as manure.
Order next season's seeds and seed potatoes.
Give stock extra rations in particularly cold periods.
Where available, heather and willow twigs are fed to sheep and goats.
Plant Jerusalem artichokes.
Sow field or tic beans as livestock feed.
Check buildings for rats and apply poisoned bait.
Put your feet up, catch up on reading, or, better still, have a winter holiday.

Spring
March–May

Harrow pasture both ways in order to pull up dead grass, leaves and moss.
Resow bare patches of grass.
Roll pasture to level and consolidate grasses lifted by the frost.
Harrow winter-ploughed soil to obtain a fine tilth.
Sow new grass ley and roll. If preferred, sow ley with a cereal crop so that by the time the cereal is harvested the grass is established underneath and, meanwhile, has had bird protection.
Sow other crops on harrowed soil. Useful winter fodder crops are: cow cabbage, carrots, kale, fodder beet, mangolds; but do not feed mangolds until January as they are slightly toxic before then.
Plant potatoes.
Sow rape for grazing in late summer.
Provide bird protection for early growth.
Begin grazing from late spring but watch out for scouring from lush grass.

Summer
June–August

Sow turnips and swedes for feeding livestock from late autumn to winter.
Haymaking June–July. Take a second cutting of hay in August if it is good enough, otherwise leave to provide 'foggage' or late grazing.
Control grazing with an electric fence to make economical use of grass.
Harvest cereal crops July–August.
Make sure that cereal crops are adequately stored and protected from rats.
Check stored hay to make sure it is free of damp.
In late summer, cut bracken if available for animal bedding, but ensure that stock is well fed when it is given so that they are not tempted to eat it. It is toxic.
Harvest potatoes and store 'chat' or small ones for livestock feeding.
Construct clamps for storing fodder root crops.
Make silage for winter use.

The Kitchen Garden

Autumn
September–November

Begin to lift maincrop potatoes. Lift and store early sown carrots and beetroot. Lift and dry onions and store in strings when ripened.
Sow winter lettuce, winter spinach and turnips for tops or fodder.
Remove celery suckers and earth up plants.
From October blanch endives.
Cut down feathery asparagus fronds.
Earth up leeks.
Store remaining maincrop potatoes in a clamp.
Clear ground where potatoes and root crops were.
Burn old potato and tomato haulms.
Make new compost heap.
Put cloches over winter lettuces, cabbages, cauliflowers and other over-wintering plants as soon as the weather worsens.
Cut fruited raspberry canes to the ground and tie in new canes which will fruit next time.
Apply thick mulch of well-rotted manure to fruit trees and fruit bushes, making sure it does not touch the stems.
Plant strawberries.

Winter
December–February

Lift Jerusalem artichokes, leeks, parsnips and celery as required. Pick sprouts, savoys, winter cabbage and spinach.
Check stored root crops for signs of decay.
Dig vacant ground and apply lime where necessary.
Plan next season's garden and order seeds and seed potatoes.
Set potatoes to 'chit' or sprout in boxes or trays.
Sow broad beans.
In February plant onion sets and garlic.
Sow tomatoes, peppers and cucumbers in heated greenhouse, for later planting.
Plant out earlier sown onions
Earth up asparagus beds and apply well-rotted manure to the surface.
Sow spring onions.
Sow parsnips.
Prune apple trees.
Plant new fruit trees and bushes.
Cut newly planted fruit bushes back to two or three buds, leaving an outward-facing bud.
Prune greenhouse vines by cutting back laterals to outward-facing bud.

Spring
March–May

Sow in succession – summer cabbage, lettuce, radish, peas, parsley, spinach, carrots.
April – cut asparagus and spring cabbage.
Pull radishes as needed.
Plant early potatoes and globe artichokes.
Stake peas.
Keep soil between plants hoed to discourage annual weeds.
Take precautions against bird attack and if necessary use netting.
Sow annual herbs.
Plant out autumn-sown herbs.
Plant greenhouse tomatoes, peppers and cucumbers.
Sow sweetcorn in greenhouse for later planting out in June or sow outside in late May.
Plant out autumn-sown rhubarb.
Sow kohl rabi.
Make a new asparagus bed.
Apply wood ashes to blackcurrant bushes as a source of potash.
Start picking early forced rhubarb.
Start using over-wintered lettuce from cloches.
Plant maincrop potatoes.

Summer
June–August

Pick broad beans, peas and spinach and freeze as needed.
Sow beetroot, carrots, turnips and lettuce.
Thin earlier sown crops.
Apply mulch between plants.
Plant out tomatoes, peppers, marrows, brussels sprouts, cauliflowers and celery.
Watch out for aphid attack.
Water as necessary.
In July start to lift early potatoes.
Sow spring cabbage, spinach, beetroot and fodder kale.
Plant leeks and late cauliflowers.
Support growing plants.
Harvest runner and French beans as they become ready.
Sow winter radish China Rose and Lamb's lettuce for pulling in the winter when other salad crops are in short supply.
Freeze, bottle, jam and pickle as harvesting proceeds.
Summer prune cordon fruits by cutting back current year's laterals to four or five leaves.
Summer prune apple trees by removing thin spindly uprights.
Sow perennial herbs and onions.

Poultry

Autumn
September–November

Allow laying hens to graze harvested arable fields and use them to clear fruit cage and greenhouse beds.
Meanwhile, prepare winter housing with electric light and covered run for winter egg production.
Cull old and non-productive hens.
Buy point-of-lay pullets for winter laying if necessary.
Give all birds a dusting of lice and mite powder such as Pybuthrin, before putting in winter quarters.
Check feet and legs for scaly leg mite attack and treat with benzyl benzoate if affected.
Worm all over-wintering stock with poultry vermifuge, by adding it to their drinking water.
Transfer breeding geese to orchard for the winter and let them eat windfalls in addition to a concentrate ration.
Put Christmas geese in a fattening area and give extra rations.
Buy Turkey poults for Easter.

Winter
December–February

Take action against rats and if necessary call on local authority help.
Give all poultry extra rations to cope with the cold.
In icy weather give warm water to drink.
Make sure supplies of crushed oystershell and limestone grit are available.
Break ice on pond to let the ducks swim.
Slaughter Christmas poultry, leaving time for the birds to hang.
Repair unused outdoor housing.
Lime pasture which poultry has vacated.
Prepare broody pens for the spring. Check incubator.
Have breeding stock blood-tested to ensure freedom from inheritable diseases.
Trim spurs of breeding cock.
Allow geese access to breeding pens from late January onwards.
Keep a wary eye open for foxes and lock up your birds.
Beware two-legged Christmas poachers.

Spring
March–May

Inject new chicks against Marek's disease. Keep in a protected environment if the weather is cold.
Buy new point-of-lay pullets if needed. Make sure they come from a reliable source and from blood-tested breeding stock.
Allow to range as soon as the weather is suitable and the grass is growing.
Buy meat-breed chicks or ducklings for rearing as table birds. Raise in protected environment.
Protect goslings from weather extremes and rats. Do not allow to graze until they are fully-feathered – and even then only if weather is mild and pasture clean.
Worm goslings once every six weeks while they are grazing. Worm any other free-ranging poultry regularly.
Fatten surplus cockerels in protected area.
Do not allow ducklings to swim in water until they are six weeks old.
Sell young stock.

Summer
June–August

Continue to allow range grazing until grass stops growing. Ensure that grass is used in rotation.
Put 'Michaelmas' or 'green' geese that are to be sold in September in a fattening area and give extra rations for three weeks before selling.
Sell surplus stock.
Buy turkey poults for rearing for Christmas.
Select any outstandingly good birds as future breeding stock and put identifying leg rings on them.
Plan a selective breeding programme and construct 'trap' nests for next season.
Continue to take action to get rid of lice and mite attack, by regular dustings.
Make sure that hens have an area which they can use for taking dust baths.

Rabbits

Autumn
September–November

If a beginner interested in table rabbits, contact the Commercial Rabbit Association. If older breeds are required, contact the British Rabbit Council.
Decide on system of housing – is it to be hutches or cages? Visit a commercial rabbitry to see stock and equipment.
Once housing is arranged, buy breeding stock in October, at 12–14 weeks old – but only if winter lighting is available. Otherwise leave until late February. Check that new stock has been vaccinated against myxomatosis.
Existing herds should be culled, with surplus being sold or going in the freezer. The small rabbit-keeper is unlikely to want to try winter breeding because of the difficulties and high feeding costs.
Read up about rabbits and, ideally, go on a management course run by C.R.A.
Keep records of all transactions and activities.

Winter
December–February

Make regular checks on ears and fur for evidence of mite infestation. For ears apply a few drops of Carbaryl or similar preparation. For fur use Pybuthrin powder. Trim nails once every six weeks.
In periods of extreme cold, extra feeding may be needed. Hay is always welcome but beware of giving too much concentrate otherwise breeding stock becomes fat and lazy, and there may be mating problems.
Where winter breeding does take place, make sure nestboxes are well insulated or young will die of cold. For mating, introduce doe into buck's cage, never the other way around.
Use manure as activator for a new compost heap, or use droppings for making liquid manure.

Spring
March–May

The main breeding period.
Feed pregnant does a maintenance ration of 113 g (4 oz) pellets a day, plus hay, for first three weeks of pregnancy. From the 21st day, double the pellet ration to cater for growing young. Where home-produced food is given, avoid too much bread and potatoes. Give a few raspberry leaves in the diet every day for the last week to aid birth.
On 27th day provide doe with a nestbox and hay or wood shavings. Kindling is between 28 and 33 days, usually on 31st. Keep careful records of all live and dead births.
If rabbits are for the table, separate from mother at four weeks and continue with *ad lib* pellet feeding, while mother reverts to maintenance ration. Weigh young ones regularly and cull at 1·8–2·2 kg (4–5 lb) at 9–12 weeks. A less intensive diet will mean later development.
Cure skins, but if there is a delay, stretch and salt them for temporary storage.
Register new, potential show rabbits with B.R.C. Leg rings must be put on before eight weeks.

Summer
June–August

Make every effort to ensure adequate ventilation in house and take action against flies.
Where summer grazing is carried out, provide stock with temporary grazing arks so that they have access to grass. Avoid areas used by other stock or fouled by cats and dogs. Keep buck separate from rest of stock.
Continue breeding through the summer and early autumn, keeping records as needed. Attend any rabbit shows in locality.
Sow fodder crops such as kale, carrots, turnips and parsley for winter and early spring feeding (the latter may need cloche protection in some areas).
Buy in or otherwise acquire hay and straw.
Give a worming preparation to any rabbits that have been grazing. Pellet-fed and cage- or hutch-housed stock should not require this. If in doubt ask the vet to do an examination of the droppings.

Bees

Autumn
September–November

Replace queens that are more than two years old, or those which are in some way unsatisfactory.
Start to feed sugar syrup from September onwards, unless there is a lot of heather nearby (2·2 kg (5 lb) white sugar to 1·4 litres ($2\frac{1}{2}$ pints) water). Aim to feed regularly until weather turns cold and bees become sleepy. Watch out for robber bees.
Make entrance hole smaller so that, although bees can get in and out, mice are excluded from taking up winter quarters.
Remove the queen excluders and supers.
Check that roof is waterproof.
In exposed positions or where high winds are likely to be a problem, make hives more stable by putting stone slabs on the roof or pegging the whole hive to the ground with ropes and pegs.
Check that ventilators are clear.
If woodpeckers are a nuisance, protect hives with garden netting.
Join local beekeeping society if not already a member.

Winter
December–February

The less interference the better.
Make periodic checks to make sure the hives have not been disturbed by winds or other weather conditions.
Continue checking that the ventilators have not become blocked.
If snow is on the ground, put a board in front of the entrance hole to stop bright reflected light from entering and perhaps tempting the bees to come out. (Do not however block the hole.)
If rain is particularly heavy place a piece of wood behind and underneath the hive so that it tilts forward slightly, making the water run off.
Check all equipment so that it is in a good state of readiness for the spring.
Order catalogues from suppliers and decide on what needs to be bought for the new season.
Plant a hedge around the hives to provide wind protection or in front of the hives as a flight director – Lawson Cypress, willow, beech and hawthorn are suitable.

Spring
March–May

Provide fresh water near hives.
If sugar feeding was inadequate in autumn, feed in March as a booster.
Make first inspection of hives in April. Check that: queen is present and laying; brood is healthy.
Remove mouse guards.
Second inspection in May: clean tops of frames; remove debris from floorboard; replace combs that are in poor condition; put on another brood box if there is not enough room for the expanding colony.
If going in for bees for the first time, acquire stock from a reputable supplier and ask an experienced beekeeper to oversee your first attempts at setting up hives and introducing stock.
Remove surplus queen cells and ensure there is enough room for colony, to prevent swarming.
Thereafter make periodic inspections when really necessary.

Summer
June–August

At end of May to beginning of June, add the first super over a queen excluder.
When bees are covering outer combs of first super, add a second super. It is in the supers that honey is laid down. Continue to add more supers as required through the summer.
If swarming occurs, take the swarm from its position (often a high branch), making sure you have a stout box. Then prepare a new hive with a brood box and foundation frames. Put a board from the ground to the entrance and cover it with a white sheet. Now place the swarm onto the sheet so that they crawl up the sheet to the hive. (Rubbing the inside of the hive with broad bean flowers is a traditional incentive for them to go in.)
Feed newly-hived swarms for a few days, until they build up their own stocks of food.
Harvest honey in late August or early September.
Remove supers, making sure that cells are capped (or honey will not be ripe). Extractors can often be hired from local beekeeping societies.

Goats

Autumn

September–November

The main breeding time. Watch for signs of sideways tail wagging, swollen vulva and persistent bleating. Take nannies to pre-arranged stud billy.
Where goats are taken off your land, or brought onto it, make the appropriate entry in a Movement of Livestock Record Book.
Give females a dusting of lice and mite powder such as Pybuthrin before and after visiting the billy.
If buying stock for the first time, choose good quality, registered goats, following advice of the British Goat Society publications.
Give worming drench such as Panacur or Thibenzole to new stock and check their feet. Trim overgrown nails and thereafter give them a file with a Surform blade once a month to keep in condition.
Give an anti-clostridial vaccination such as Covexine.
Buy good quality milking equipment – if not already available. Read up about milk hygiene – and buy yourself an overall.

Winter

December–February

Keep goats inside if wet or particularly cold. Ensure that adequate hay and concentrates are given to counteract effects of severe cold, and also to cater for needs of developing kids. Give warm water to drink on very cold days.
Feed winter fodder such as kale where available, but do not overfeed in case of bloat. Ivy leaves and twigs (but not berries which are toxic) are welcomed, as well as willow branches.
Give extra straw as bedding in cold weather, to provide added insulation. Ensure that bedding is added regularly so that damp areas do not form.
Continue with regular feet checks and nail filing.
Ensure that goats have access to a mineral lick.
Feed supplements where needed.
Worm in-kid goats six weeks before kidding.
Check First Aid Box and add to as necessary.
Catch up on reading and join local goat society if not already a member.

Spring

March–May

The main kidding time. Have the vet's telephone number to hand in case of emergencies. Ensure that dried milk supplies, bottles and teats for bottle feeding of kids are ordered and delivered in good time before kidding.
Have kids disbudded during the first week so that horns do not develop.
Allow kids to feed from the mother for four days to ensure that they receive colostrum, before removing them to start bottle feeding. Keep the kids together in their own pen so that they have company.
Castrate male kids if they are being reared for meat; otherwise have them put down.
Have new kids earmarked through auspices of local goat societies.
Register kids with British Goat Society.
Give all stock an anti-clostridial injection.
Watch for signs of scouring from early lush grass.
Be prepared for lots of extra milk. Order yoghurt and cheese cultures and dairying supplies.

Summer

June–August

Continue to freeze and date surplus milk supplies so that there is no shortage of milk in the winter.
Obtain hay supplies and store in sheltered, dry conditions.
Continue worming stock once every six weeks until October.
Check regularly for signs of lice and mites and ticks which are particularly prevalent in summer.
Sell kids and any other stock not being taken through the winter.
Buy in adequate supplies of straw as bedding, and store in dry conditions.
Keep a watch out for signs of mastitis in any goat being dried off.
Slaughter male kids being raised for meat unless the weather is hot, in which case delay until cooler.
Obtain stud list from local goat society and arrange a visit as far as is approximately possible.
Arrange transport in good time.

Cattle

Autumn

September–November

Treat all cattle with a prescribed skin wash for Warble fly between September and October – no later.
Check feet regularly and pick out hooves.
Trim and rasp hooves regularly.
If acquiring stock for the first time, get advice from A.D.A.S. and the Milk Marketing Board, as well as the individual breed societies.
Dry off pregnant cows three months before calving.
Watch out for signs of mastitis.
Worm all stock.
Sell any stock which is not being taken through the winter.
Slaughter bull calf which has been raised for the freezer; choose a cold day.
Sell any autumn born calves as fatstock, unless being kept as additions to the dairy herd.

Winter

December–February

Provide sheltered housing for cattle during the worst of the winter.
Prepare sheltered area for cows due to calve.
Feed silage and fodder crops such as kale, fodder beet, swedes and mangolds, in addition to hay and concentrates.
Let new calves feed on mothers for four days, then bucket feed.
Eartag calves with herd number or ear tattoo if registered with a specific breed society by two weeks old.
Dehorn in first few days.
Castrate bull calves if necessary.
Lime pasture where necessary.
Spread manure or feed nitrogen to pasture – but not at same time as lime.
Check and repair fences.
Cut hedges.
Drain ditches.
Prepare ground by ploughing for any spring-sown grass leys or fodder crops.
Sow field beans such as Maris Beagle.
Carry out regular checks and maintenance of any milking equipment.

Spring

March–May

Let calves go out to graze from May onwards.
Wean onto solids completely at 8–12 weeks.
Watch out for signs of scouring or magnesium deficiency – 'grass staggers' from new lush grass.
Provide mineral licks.
Worm all stock.
Have 'bulling' heifers of about 15 months mated by A.I. in order to calve early in the year. Otherwise choose time of year which is best suited to your particular needs. A.D.A.S and M.M.B. will advise on availability of A.I.
Control grazing with electric fencing.
Sow fodder beet and carrots.
Sow kale in succession: Marrowstem – for use before end of December; Thousand-Headed – for December/January; Hungry Gap – for February/March; Maris Kestrel – until March.
Sow new grass ley.
Sow herbal ley for strip grazing, e.g. a mixture of ribgrass, trefoil, yarrow, white clover, sheep's parsley, burnet and chicory.
Select future additions to milk herd from calves.

Summer

June–August

Haymaking in June.
Keep a watch out for signs of lice, mite and tick attack.
Sow fodder crops – kale, cow cabbage, swedes and mangolds.
Control grazing with electric fencing.
Cut grass for silage making.
Keep bullocks and stock being raised for beef separate from dairy herd. Give best pasture to milking animals.
Harvest field beans.
Acquire straw supplies for bedding.
Acquire barley straw for winter feeding of bullocks which are being taken through to a second year.
Sow tares for use either as a late grazing crop or for cutting for silage.
Make regular checks on grass to make sure that poisonous plants such as ragwort do not become established.

Sheep

Autumn
September–November

If forming a flock buy breeding ewes in September. Check whether they have been injected against clostridial diseases. If not, inject with a vaccine such as Tasvax 7, or one recommended by the vet. Check the ewes all over, and especially the teeth and feet. Teeth are an indication of age and 'broken-mouthed' ewes should be avoided. Feet should be checked for footrot and trimmed if nails are overgrown.
Run the flock through a foot bath containing a 10 per cent copper sulphate solution, or a 6 per cent formalin solution as a precaution against footrot. Do this whenever the flock is being moved for any reason.
'Flush' the ewes – bring to peak condition before breeding – by giving them access to good pasture.
Dip against sheep scab about one week before breeding.
Mating occurs in October. Let ewes run with ram fitted with a sire harness and dye.
Give second clostridial dose to new stock.

Winter
December–February

Let pregnant ewes forage and generally clear up fields and forage crops, but do not let them over-eat. Electric fencing is the best control.
As the weather worsens, bring ewes to a sheltered area. Feed hay to replace grass. Willow branches are also appreciated.
From January onwards, give pregnant ewes a concentrate ration to cater for the developing lambs. (This is generally referred to as 'steaming up' or getting them into good condition for lambing.)
Give ewes a vermifuge such as Panacur or Thibenzole, five to six weeks before lambing so that parasitic worms are killed.
Prepare lambing pens in February.
Vaccinate ewes against clostridial diseases so that unborn lambs have protection.
Enquire whether there are any local lambing courses run by the Ministry of Agriculture – they are excellent. Contact local A.D.A.S. office via Yellow Pages.

Spring
March–May

March – the main lambing period. Be prepared for some sleepless nights. Have the telephone number of the vet or experienced sheep handler to hand.
Make sure dried milk supplies such as Ewlac, bottles and lamb teats for orphan lamb feeding have been ordered and received in good time.
Ensure that lactating ewes are being adequately fed, and make a small amount of concentrates available to the new lambs ('creep' feeding).
Castrate ram lambs.
Dock tails of lambs.
Earmark lambs if necessary.
Keep a careful eye open for signs of scouring from lush early grass.
Worm ewes and check feet.
Run them through a footbath.
Worm lambs in April.
In May, vaccinate lambs against clostridial diseases – except for those which are being sold as early meat lambs.
Make sure that sheep have access to mineral licks for adequate mineral supplies.

Summer
June–August

June – shear sheep.
Give lambs a second dose of anti-clostridial vaccine.
Dip sheared sheep against external parasites such as ticks, mites, blowfly larvae and lice.
July – wean lambs.
Keep a careful eye on the ewes during this period in case mastitis develops.
Give ewes a booster vaccination against clostridial diseases.
Worm ewes and lambs.
Sell fat lambs – lambs which have been kept until they are heavier and command a better price later in the year. Where lambs are being kept for the home freezer, give them some good pasture to 'finish them off' before they are slaughtered in September.
Sell surplus breeding stock in August or early autumn.

Pigs

Autumn
September–November

Acquire breeding stock if starting a pig unit. Unless going in for older breeds, avoid having a boar; use A.I. instead.
Check that breeding gilts are from herd that is free of endemic enzootic pneumonia and that they have been wormed.
Check feet and trim if necessary.
Make sure that the necessary permit to move pigs has been applied for in time, and that the appropriate entry is made in the Movement of Livestock Record Book.
Do not breed in autumn, unless geared to do so with appropriate buildings, etc., or young will be born in winter. Leave until November at the earliest so that births are from April onwards.
Let breeding stock forage in orchards and clear up arable and other fields, so that they are lean and hardy before mating.
Slaughter any porkers or baconers that have been raised for the freezer; the colder the weather the better.

Winter
December–February

Let breeding pigs plough land for cultivation in spring. Control with electric fencing. In severe weather move to more protected areas and housing.
Give extra rations where needed, to counteract effects of cold, but do not overfeed. A dry sow will need approximately 1·8 kg (4 lb) of meal a day if indoors.
Ensure that fodder crops such as turnips and swedes are well chopped before feeding.
Watch for signs of heat in non-pregnant sows and arrange for A.I. as soon as possible.
Ensure pregnant sows have adequate protein in their diet.
Prepare farrowing house with strawed floor, crush bar to make safe area for piglets, and an infra-red lamp.
Worm sows about six weeks before farrowing.
Treat for lice and mites if there is any indication of infestation.

Spring
March–May

Wash sow's udder before farrowing.
Be with her during farrowing and place piglets in warm, protected place until all are born.
Feed lactating sows 1·3 kg (3 lb) meal plus 0·4 kg (1 lb) for each piglet, per day.
Give iron injections to indoor-farrowed piglets.
Earmark if necessary.
Castrate boar piglets if necessary.
Give 'creep' feed to piglets from the age of two weeks (small amounts of food in their own protected area).
Wean at about seven weeks and give weaner meal.
Watch out for mastitis in sows.
Worm piglets two weeks after weaning.
Select future breeding stock from piglets.
Sell surplus weaners as pork pigs.
If starting with pigs at this time of year, buy weaners from a healthy and reputable source and fatten to pork weight – 45–68 kg (100–150 lb) – keeping the best ones to grow on and provide breeding stock for next year.

Summer
June–August

Sell any pigs kept on after pork weight as baconers – 73–91 kg (160–200 lb).
Ensure that outside pigs have a frequent change of pasture in order to prevent a build-up of parasitic worms.
Worm sows.
Make sure that light pigs have adequate shelter from the sun if they are outside in hot weather (they can get sunburn).
Clean out, fumigate and generally repair any buildings not in current use, e.g. winter housing, etc.
Sow fodder crops such as kale, turnips, swedes, beet and mangolds.
Acquire straw supplies for bedding for the year.
Watch out for signs of lice and mite attack and treat with a veterinary insecticide such as Pybuthrin.
Sell surplus breeding stock.

APPENDIX: Coping with problems the organic way

There are people who claim that it is impossible to grow vegetables properly without chemicals. Our experience, like that of many others, is that we can. The following table lists some of the ways in which it is possible to avoid problems and to cope with them when they do come. If you have not already come across the excellent organization The Henry Doubleday Research Association, it is the society for Britain's organic gardeners. The address is: H.D.R.A., 20 Convent Lane, Bocking, Braintree, Essex.

Problem	**Plants affected**	**What to do about it**
Slugs	Most vegetables, but particularly tender ones and those under cloches	Let ducks or chickens clear up kitchen garden in autumn in order to reduce population available to breed. (Protect winter vegetables such as Brussels sprouts, or they will be eaten too.) Place empty gratefruit halves on soil and leave overnight to attract slugs. Put a little stale beer in an old saucer sunk into the ground so that they drown in bliss. Try a non-poisonous product such as Fertosan, available in garden retailers.
Blackfly (aphids)	Many plants, but a particular nuisance on broad beans	Sow early enough to avoid attack. Pinch out tender growing tops after rest of plant has produced flowers, to make them less attractive. Spray with dilute seaweed extract as a deterrent. If blackfly already apparent, spray with salt water.
Woolly Aphis (American blight)	Fruit trees	Dab the woolly-looking patches (which contain aphids) with methylated spirits.
Cabbage root fly	All brassica family	Place protective collars around stems of plants (see kitchen garden section for details). If affected, dissolve 5 ml (1 tsp) of disinfectant in 4·5 litres (1 gallon) of water onto soil around plant stems.
Clubroot	Brassica family	Ensure that plot is well limed and sprinkle extra circle of lime around each plant. Always grow your own brassica plants from seed. Never buy in from elsewhere.
Carrot fly	Carrots	Concentrate on early crops. Avoid brushing foliage as much as possible, for the scent attracts the fly. Thin out seedlings in wet weather. Sprinkle salt on soil around carrots as a deterrent. If affected, water disinfectant solution, as detailed for cabbage root fly. Interplant carrots with onions.
Onion fly	Onion family	Grow onions from 'sets' instead of seeds which are more likely to be affected. Interplant with carrots. If affected, water soil around onions with disinfectant solution (as for cabbage root fly).
Caterpillars	Brassicas, gooseberries and many other plants	If you see them, pick them off by hand (it's not that big a task) and feed to the ducks. Spray with salt water.
White fly	Tomatoes, cucumbers – a particular nuisance in greenhouses	Do what the commercial growers are doing and use a biological control in the form of the mini-wasp, *Encarsia formosa*, which is parasitic on whitefly.
Red spider mite	Greenhouse crops	Again, try the biological control of using the predator, *Phytoseiulus persimilis*.

GLOSSARY

Ad-lib feeding: allowing livestock to feed when they want to.

Annual: a plant which completes its life cycle in one season.

Antibiotic: a medication to treat bacterial infections.

Aspergillosis: a fungal disease affecting the lungs of livestock and humans, caused by spores in mouldy hay. The common name is 'farmer's lung'.

Anti-bodies: naturally produced substances in the body which help to counteract infection.

Bantam: a miniature or scaled-down version of a poultry breed.

Biennial: a plant which completes its life cycle in two years.

Billy: a male goat.

Blanching: the process of excluding light from plants in order to whiten them, e.g. drawing up soil around leeks; *or:* the process of subjecting vegetables to boiling water or steam before they are frozen.

Boar: a male pig.

Broiler: a fowl bred specifically for the table.

Broiler ration: a proprietary ration formulated for the needs of table birds.

Brooding: the provision of protected conditions for chicks in the vulnerable period after hatching.

Buck: a male rabbit or male goat.

Bullock: a young, unmated or castrated bull.

Bulls: male cattle.

Calves: young cattle up to the age of 6 months.

Calving: giving birth to calves.

Chick: a young fowl up to the age of seven weeks.

Chicken: a young fowl of either sex, up to the age of 12 months.

Clamp: a protected area of straw and earth, for keeping root crops over the winter.

Cock: a male fowl over the age of 12 months.

Cockerel: a young male fowl up to the age of 12 months.

Colostrum: the first milk produced after birth which is rich in nutrients and antibodies, for the protection of the young animal.

Cows: female cattle.

Dairy ration: a proprietary feed specially formulated for the needs of milk producing livestock.

Doe: a female rabbit or female goat.

Drake: a male duck.

Duckling: a young duck up to the age of seven weeks.

Farrowing: giving birth to piglets.

Flushing: bringing up to peak condition for mating.

Fodder: fibrous food for livestock.

Gander: a male goose.

Gestation: period of pregnancy.

Gilt: a young, unmated female pig.

Goatling: a young goat between three and 12 months.

Gosling: the young of geese, up to the age of three months.

Grinding: the process of crushing grain.

Hay: grass which is preserved by drying.

Heifer: a young, unmated cow.

Hen: a female fowl over the age of 12 months.

Incubation: the process of development of the chick embryo while in the shell.

Kibbling: the process of chopping grain.

Kid: a young goat up to the age of three months.

Kidding: giving birth to goat kids.

Kindling: giving birth to rabbits.

Lactation: milk production.

Lambing: giving birth to lambs.

Layer: a fowl bred specifically for egg production.

Layers' pellets: a proprietary poultry ration specially formulated for the needs of laying birds.

Maincrop: a crop which is sown or planted later than the first early sowing and provides the main crop of a particular variety.

Mulch: a layer of material such as peat, grass

mowings, paper or plastic, placed on soil in order to conserve moisture and suppress weeds.

Nanny: a female goat.

Notifiable disease: a condition which must be notified to the local authorities e.g. fowl pest, rabies.

Pectin: a natural setting or solidifying agent in fruit.

Perennial: a plant which grows every year.

pH value: the degree of acidity or alkalinity of a soil.

Piglet: a young pig up to the age of six weeks.

'*Pop-hole*': a small exit which can be open or closed to allow domestic birds entry to or exit from a poultry house.

Poult: a young turkey, up to the age of three months.

Pullet: a young female chicken before the onset of laying.

Rooster: an American name for a cock.

Scouring: a digestive upset giving rise to diarrhoea.

Silage: grass which is preserved by a 'pickling' process to provide winter feed for livestock.

Soft fruit: fruit that comes from shrubs or herbaceous plants, e.g. blackcurrants, gooseberries, strawberries.

Sow: a female pig after first mating.

Stag: a male turkey.

Staggered planting: a system of planting where plants are placed according to a diamond pattern, rather than in rows, to make best use of space.

Straw: the dried stems from a cereal crop.

Top fruit: fruit which comes from trees rather than shrubs, e.g. apples, pears, plums.

Tupping: the mating of sheep.

Vermifuge: a medication to kill internal, parasitic worms.

Weaner: a young animal which has been weaned from a milk diet onto solid foods.

USEFUL ADDRESSES

United Kingdom

MINISTRY OF AGRICULTURE AND FISHERIES (Agricultural Development and Advisory Service), Great Westminster House, Horseferry Road, London, S.W.1. (See local Yellow Pages Guide for address of nearest divisional office.)
BRITISH BEEKEEPERS' ASSOCIATION, Hightrees, Dean Lane, Merstham, Surrey.
BRITISH GOAT SOCIETY, Rougham, Bury St Edmunds, Suffolk.
BRITISH POULTRY FEDERATION, 52–54 High Holborn, London, W.C.1.
BRITISH WATERFOWL ASSOCIATION, c/o Market Place, Haltwhistle, Northumberland.
COMMERCIAL RABBIT ASSOCIATION, Tyning House, Shurdington, Cheltenham, Gloucestershire, GL51 5XF.
THE HENRY DOUBLEDAY RESEARCH ASSOCIATION, 20 Convent Lane, Bocking, Braintree, Essex.
THE JERSEY CATTLE SOCIETY OF GREAT BRITAIN, 154 Castle Hill, Reading, Berkshire.
NATIONAL PIG BREEDERS' ASSOCIATION, 7 Rickmansworth Road, Watford, Hertfordshire.
NATIONAL SHEEP ASSOCIATION, Jenkins Lane, St Leonards, Tring, Hertfordshire.
ORGANIC GROWERS' ASSOCIATION, Aeron Park, Llangeitho, Tregaron, Dyfed.
HOME FARM (formerly *Practical Self Sufficiency*) Broad Leys Publishing Company, Widdington, Saffron Walden, Essex, CB11 3SP – Britain's smallholding magazine.
SELF-SUFFICIENCY AND SMALLHOLDING SUPPLIES, The Old Palace, Wells, Somerset – tools and equipment.

United States of America

UNITED STATES DEPARTMENT OF AGRICULTURE (U.S.D.A.), Washington D.C., 20505.
AMERICAN BEE SOCIETY, Hamilton IL. 62341.
THE AMERICAN DAIRY GOAT SOCIETY, Box 186, Spindale NC. 28160.
THE AMERICAN JERSEY CATTLE CLUB, 2105 J. South Hamilton Road, Columbus OH. 43227.
THE AMERICAN POULTRY ASSOCIATION, Box 70, Cushing OK. 74023.
AMERICAN RABBIT BREEDERS' ASSOCIATION, 1007 Morrissey Drive, Bloomington IL. 61701.
COUNTRYSIDE AND SMALL STOCK JOURNAL, Rt. 1. Box 239, Waterloo WI. 53594.
INTERNATIONAL WATERFOWL BREEDERS' ASSOCIATION, 12402 Curtis Road, Grass Lake MI. 49240.
MOTHER EARTH NEWS, Box 70, Hendersonville NC. 28739 – small farming magazine.

Canada

MINISTRY OF AGRICULTURE, 930 Carling Avenue, Ottowa, Ontario.
CANADIAN JERSEY CATTLE CLUB, 343 Waterloo Avenue, Guelph, Ontario.
CANADIAN ORGANIC GROWERS, 33 Karnwood Drive, Scarborough, Ontario.
HARROWSMITH, Camden House Publishing Ltd., Camden East, Ontario – small farming magazine.
HOMESTEAD EQUIPMENT, Box 339, Acton, Ontario – tools and equipment.
ONTARIO DAIRY GOAT SOCIETY, RR3, Crysler, Ontario.
THE PIONEER PLACE, Box 100, Route 4, Aylmer, Ontario – supplies and equipment.

Australia

DEPARTMENT OF AGRICULTURE AND FISHERIES (DAF), 25 Grenfell Street, Adelaide 5000.
EARTH GARDEN, PO Box 378, Epping, NSW, 2121 – small farming magazine.
GOAT BREEDERS' SOCIETY OF AUSTRALIA, Box 4317, GPO Sydney, NSW, 2001.
GRASS ROOTS, Box 900, Shepparton 3630 – small farming magazine.
ORGANIC GARDENING AND FARMING SOCIETY OF TASMANIA INC., GPO Box 228, Ulverstone 7315.
SELF-SUFFICIENCY SUPPLIES, 256 Darby Street, Newcastle 2300 – supplies and equipment.

BIBLIOGRAPHY

Part-Time Farming, Katie Thear, Ward Lock 1982. Concentrates on the commercial aspect of making an income from a small farm

Organic Gardening, Lawrence D. Hills, Penguin 1981. A coverage of gardening without chemicals

A Modern Herbal, Grieve & Leyel, Penguin Handbooks 1982. The most comprehensive herbal available

The Fruit Finder, Lawrence D. Hills, Henry Doubleday Research Association 1976.

The Fruit Garden Displayed, Royal Horticultural Society 1975

Keeping Chickens, Waltens & Parker, Pelham Books 1982

Raising the Home Duck Flock, D. Holderread, Garden Way 1978

Keeping Domestic Geese, Barbara Soames, Blandford 1980

Raising Your Own Turkeys, Leonard Mercia, Garden Way 1981

Practical Rabbit-Keeping, Katie Thear, Ward Lock 1981

The Complete Handbook of Beekeeping, Herbert Mace, Ward Lock 1976

Goat Husbandry, David Mackenzie, Faber 1980

The Home Dairying Book, Katie Thear, Broad Leys Publishing 1978

Beef Management and Production, Derek H. Goodwin, Hutchinson 1971

The Book of the Pig, Susan Holme, Saiga Publishing 1979

Sheep Management and Production, Derek H. Goodwin, Hutchinson 1979

Home Preservation of Fruit and Vegetables, H.M.S.O.

Keeping Warm for Half the Cost, Colesby & Townsend, Prism Press 1981

Black's Veterinary Dictionary, Geoffrey West, Adam & Charles Black 1982

Home Farm (formerly *Practical Self Sufficiency*) magazine. Broad Leys Publishing Ltd, Widdington, Saffron Walden, Essex CB11 3SP

INDEX